Disclaimer

The publisher of this book is by no way associated with the National Institute of Standards and Technology (NIST). The NIST did not publish this book. It was published by 50 page publications under the public domain license.

Book Title: A Threat Analysis on UOCAVA Voting Systems

Book Author: Andrew R. Regenscheid; Nelson E. Hastings

Book Abstract: This report contains the results of NIST s research into technologies to improve the voting process for United States citizens living overseas. It splits the overseas voting process into three stages: voter registration and ballot request, blank ballot delivery, and voted ballot return. For each stage, this report describes how various transmission options could be used to support overseas voting. The transmission options discussed in this paper are postal mail, telephone, fax, electronic mail, and web-based systems. This report documents a threat analysis based on the methodology provided in NIST SP 800-30 for each method. As part of the analysis, mitigating controls for each threat are provided when possible. The mitigating controls for each threat provided in this report provide the basis for an effort to develop best practices for overseas voting systems.

Citation: NIST Interagency/Internal Report (NISTIR) - 7551

Keyword: voting; overseas voting; UOCAVA; threat analysis; security controls

NISTIR 7551

A Threat Analysis on UOCAVA Voting Systems

Andrew Regenscheid
Nelson Hastings

**National Institute of
Standards and Technology**
U.S. Department of Commerce

[This page intentionally left blank.]

NISTIR 7551

A Threat Analysis on UOCAVA Voting Systems

Andrew Regenscheid
Nelson Hastings
Information Technology Laboratory
National Institute of Standards and Technology
Gaithersburg, MD 20899-8930

December 2008

U.S. Department of Commerce
Carlos M. Gutierrez, Secretary

National Institute of Standards and Technology
Patrick D. Gallagher, Deputy Director

[This page intentionally left blank.]

A Threat Analysis on UOCAVA Voting Systems

TABLE OF CONTENTS

EXECUTIVE SUMMARY 1
1   INTRODUCTION... 3
    1.1   SCOPE 3
    1.2   STRUCTURE OF THIS PAPER 3
2   BACKGROUND 4
    2.1   UOCAVA VOTING PROGRAMS 4
        2.1.1   *FWAB* 4
        2.1.2   *Electronic Transmission Service* 4
        2.1.3   *Voting over the Internet* 5
        2.1.4   *SERVE* 5
        2.1.5   *Interim Voting Assistance System* 6
    2.2   CURRENT UOCAVA VOTING PROCESS 7
    2.3   DIFFICULTIES IN THE CURRENT UOCAVA VOTING PROCESS 9
3   UOCAVA VOTING PROCESS 10
    3.1   VOTER REGISTRATION AND BALLOT REQUEST 11
    3.2   BALLOT DELIVERY 11
    3.3   BALLOT RETURN 12
4   DESCRIPTION OF TRANSMISSIONS OPTIONS 13
    4.1   TRANSMISSION OPTIONS 13
        4.1.1   *Postal Mail* 13
        4.1.2   *Telephone* 13
        4.1.3   *Fax* 13
        4.1.4   *Electronic Mail* 14
        4.1.5   *Web-Based* 14
    4.2   OPTIONS FOR VOTER REGISTRATION AND BALLOT REQUEST 15
        4.2.1   *Postal Mail* 15
        4.2.2   *Telephone* 16
        4.2.3   *Fax* 16
        4.2.4   *Electronic Mail* 16
        4.2.5   *Web-Based* 17
    4.3   OPTIONS FOR BALLOT DELIVERY 17
        4.3.1   *Postal Mail* 17
        4.3.2   *Telephone* 17
        4.3.3   *Fax* 18
        4.3.4   *Electronic Mail* 18
        4.3.5   *Web-Based* 19
    4.4   OPTIONS FOR BALLOT RETURN 19
        4.4.1   *Postal Mail* 19
        4.4.2   *Telephone* 20
        4.4.3   *Fax* 20
        4.4.4   *Electronic Mail* 21
        4.4.5   *Web-Based* 21

5 THREAT ANALYSIS METHODOLOGY 23
5.1 THREATS 23
5.2 THREAT SOURCES 23
5.3 EFFORT 25
5.4 DETECTION 25
5.5 IMPACT 25
5.6 POSSIBLE CONTROLS 26

6 THREAT ANALYSIS 27
6.1 REGISTRATION AND BALLOT REQUEST 27
6.1.1 Postal Mail 27
6.1.2 Telephone 28
6.1.3 Fax 29
6.1.4 Electronic Mail 30
6.1.5 Web-Based 32
6.2 BALLOT DISTRIBUTION 34
6.2.1 Postal Mail 34
6.2.2 Fax 35
6.2.3 Electronic Mail 36
6.2.4 Web-Based 37
6.3 BALLOT RETURN 39
6.3.1 Postal Mail 39
6.3.2 Telephone 40
6.3.3 Fax 41
6.3.4 Electronic Mail 42
6.3.5 Web-Based 45

7 SECURITY CONTROLS 47
7.1 POSTAL MAIL 48
7.2 TELEPHONE TRANSMISSION 51
7.3 FAX TRANSMISSION 55
7.4 E-MAIL TRANSMISSION 58
7.5 WEB-BASED TRANSMISSION 63

8 CONCLUSIONS 67
8.1 REGISTRATION AND BLANK BALLOT REQUEST 67
8.2 DELIVERY OF BLANK BALLOTS 67
8.3 RETURN OF VOTED BALLOTS 68
8.4 SUGGESTED NEXT STEPS 69

REFERENCES 70

APPENDIX: ACRONYMS 72

Executive Summary

The Election Assistance Commission (EAC), with the assistance of the National Institute of Standards and Technology (NIST), is researching electronic technologies that may help to assist overseas voting as defined by the Uniformed and Overseas Citizens Absentee Voting Act (UOCAVA). This report contains the results of NIST's research.

Uniformed and Overseas Citizens Absentee Voting Act (UOCAVA)
In 1986, Congress enacted UOCAVA, which states that U.S. citizens that are part of the uniformed services, merchant marines, and their families or citizens residing overseas are allowed to register and vote absentee for Federal office. Additionally, the Help America Vote Act of 2002 (HAVA) requires the EAC to study overseas voting, including methods for sending balloting materials to overseas voters [28]. Most states have their own legislation covering how UOCAVA citizens register and vote. Overseas voting is treated by most jurisdictions as absentee voting, applying the same procedures (e.g., deadlines for requesting absentee ballots and returning completed ballots) as for an absentee voter within the United States.

Purpose of Report
UOCAVA voting generally relies upon postal and military mail as the mechanism to distribute and receive election materials, but inherent delays in the delivery times to citizens overseas plus legislated windows of time between finalization of ballots and the election can result in UOCAVA voters being unable to participate in elections. This report therefore examines electronic transmission options (telephone, fax, e-mail, web) for UOCAVA voting that are in limited use or have been proposed as methods for improving UOCAVA voting, and analyzes the security-related threats to these transmission options. This report presents initial conclusions regarding the use of these electronic technologies and suggested next steps.

This report identifies issues and threats associated with transmitting election information by postal mail and the four electronic transmission options identified below:

- *Telephone* allows instant two-way communication between two users. Voter information can be communicated over the telephone network to or from the UOCAVA voter either verbally or by using the telephone keypad. For example, a voter could request election material by following a series of voice prompts and pressing numbers on the keypad.

- *Fax* allows users to transmit written or printed information to another party. Voter information can be scanned and transmitted over telephone networks to or from the UOCAVA voter. In some states, faxes are used as an alternative to postal mail, allowing voters or election officials to fax election forms or ballots to the other party. For example, an election official could fax a blank ballot to the fax number provided by the UOCAVA voter.

- *Electronic mail* (e-mail) allows users to send text and/or files from one computer to another over the Internet. Voter information could be sent as an e-mail message or as an attachment to the e-mail. For example, blank ballots could be sent as PDF files attached to an e-mail.

- *Web-based voting* allows users to communicate by using websites accessible via the Internet. Voter information can be presented, downloaded, or transmitted by the UOCAVA voter through the use of web pages and interactive forms. For example, voters could download blank ballots from a web site.

Initial Conclusions

The report looks at three UOCAVA election functions:
- registration and ballot request,
- blank ballot distribution to overseas voters, and
- voted ballot return.

Registration and ballot request: Voter registration and requests for a blank ballot by the UOCAVA voter can be reliably facilitated and expedited by the use of any of the electronic transmission options. The associated threats can be mitigated through the use of procedural and technical security controls and do not pose significant risks to the integrity of elections. It should be noted that e-mail and the web present greater security challenges (similar to those encountered by e-commerce applications) than telephone and fax.

Blank ballot distribution: Distribution of blank ballots to the UOCAVA voter can be reliably facilitated and expedited by the use of fax, e-mail, or web transmission. The threats associated with using fax, e-mail, and web transmission can be mitigated through the use of procedural and technical security controls and therefore do not pose significant risks to the integrity of elections. (Telephone solely to deliver blank ballots is not considered in this report as a viable transmission option for blank ballot distribution.)

Voted ballot return: Sending completed ballots from UOCAVA voters to local election officials can be expedited through the use of the electronic transmission options. However, their use can present significant challenges to the integrity of the election. Use of fax poses the fewest challenges, however fax offers limited protection for voter privacy. While the threats to telephone, e-mail, and web can be mitigated through the use of procedural and technical security controls, they are still more serious and challenging to overcome.

Recommended Next Steps

A number of states already distribute blank ballots via fax or e-mail. However, at this time there are no guidelines documenting best practices for fax, e-mail or web distribution of ballots. Developing a best practices document could help improve methods for distributing ballots using these transmission methods, and potentially improve the procedures and technical controls already in place in states currently using these methods. In addition, registration and ballot requests can also take advantage of these distribution methods, but there are more threats when handling personal information from voters. Voted ballot return remains a more difficult issue to address, however emerging trends and developments in this area should continue to be studied and monitored.

1 Introduction

The Election Assistance Commission (EAC) requested that the National Institute of Standards and Technology (NIST) research technologies to enable uniformed and overseas United States citizens to vote, as required by the Uniformed and Overseas Citizens Absentee Voting Act (UOCAVA) [21]. Additionally, the Help America Vote Act of 2002 (HAVA) requires the EAC to study overseas voting, including methods for sending balloting materials to overseas voters [28]. This report contains the results of NIST's research into technologies to enable overseas voting by United States citizens.

1.1 Scope

A general overseas voting process model was developed based on current UOCAVA practices. This report identifies three stages to the overseas voting process: voter registration and ballot request, blank ballot delivery, and voted ballot return. It describes the processes in each stage, the types of information transmitted, and the security needs for that information. In addition, a discussion of the current technologies that could be used to transmit voting information between voters and election officials is provided. Using the overseas voting process model and current technologies for transmitting voting information between voters and election officials, NIST has developed a threat analysis based on the methodology found in NIST Special Publication (SP) 800-30 *Risk Management Guide for Information Technology Systems* [2]. As part of the threat analysis, mitigating controls for each threat are provided when possible. The mitigating controls for each threat provided in this report provide the basis for an effort to develop best practices for overseas voting systems, but do not represent a set of complete and testable requirements for overseas or remote voting systems.

1.2 Structure of this Paper

The remainder of this paper is organized as follows:
- **Section 2** outlines historical and current approaches for UOCAVA voting.
- **Section 3** describes the three stages of the UOCAVA voting process: Voter Registration and Ballot Request, Ballot Delivery, and Ballot Return.
- **Section 4** identifies five transmission options for election materials: postal mail, telephone, fax, electronic mail and web-based systems. Each option is described and a typical usage scenario is provided for UOCAVA election systems.
- **Section 5** describes the threat analysis methodology used in this paper.
- **Section 6** provides the results of the threat analysis on UOCAVA election systems using the transmission options identified in Section 4 to support the three stages in UOCAVA voting.
- **Section 7** describes security controls discussed in NIST SP 800-53, Recommended Security Controls for Federal Information Systems, which can mitigate some of the threats identified in Section 6.
- **Section 8** offers conclusions based on the results from the threat analysis.

2 Background

In 1986, Congress enacted Uniformed and Overseas Citizens Absentee Voting Act (UOCAVA) [21]. UOCAVA states United States citizens that are part of the uniformed services, merchant marines, and their families, or U.S. citizens residing overseas are allowed to register and vote absentee for Federal offices. For state and local elections, most states have state legislation covering how UOCAVA citizens register and vote absentee. UOCAVA stated that a Presidential designee should carry out the mandates specified in the legislation. On June 8, 1988, Executive Order 12642 "Designation of Secretary of Defense as Presidential Designee" assigned the Secretary of Defense the administrative responsibilities for UOCAVA. In turn, the Secretary of Defense assigned these responsibilities for implementing the UOCAVA to the Federal Voting Assistance Program (FVAP) within the Department of Defense (DoD).

2.1 UOCAVA Voting Programs

2.1.1 FWAB

UOCAVA [20, 21] calls for a Federal Write-In Absentee Ballot (FWAB) covering elections for Federal offices (e.g., President/Vice President, U.S. Senator, and U.S. Representative). In addition to the FWAB, UOCAVA describes a Federal Post Card Application (FPCA) that allows citizens to request an absentee ballot for a federal election. The FVAP has made the FWAB and FPCA available at locations around the world including military bases, embassies, consulates, election organizations, and corporations as well as online electronically at their website [19]. In addition to distributing the FWAB and FPCA, the FVAP has conducted pilot projects to investigate using electronic means, such as email and websites, to assist uniformed and overseas citizens to vote.

2.1.2 Electronic Transmission Service

In 1990 as part of Operation Desert Shield, the FVAP established the Electronic Transmission Service (ETS) that allowed voters to request and receive blank ballots from their state/jurisdiction via fax as well as to return the completed ballot to their state/jurisdiction via fax. The FVAP would receive the faxed voting material (absentee ballot requests, blank absentee ballots, completed absentee ballots, etc.) from the state/jurisdiction or voter. The FVAP would forward the voting material they receive to the appropriate state/jurisdiction or voter by fax. In October 2003, the FVAP expanded ETS to include a fax-to-email conversion capability. The fax-to-email conversion capability was added to support uniformed service members serving in Iraq and Afghanistan where faxing support was limited and email support was a viable alternative. A state/jurisdiction would have to consent to use the fax-to-email conversion capability as a method to distribute voting information between the state/jurisdiction and voter. For the fax-to-email conversion, a state/jurisdiction would fax voting material to the FVAP. The FVAP would convert the voting material received by fax into a read-only PDF file that would be emailed to the voter as an attachment. The voter would print the voting material including the blank absentee ballot, complete the absentee ballot, scan the completed absentee ballot into a PDF file, and email the completed absentee ballot to the FVAP as an attachment. The FVAP would then convert the voter's PDF file into a fax for transmission to the voter's State/jurisdiction. Today, the FVAP also provides the capability to distribute voting material completely via email. Whether a completed absentee ballot is returned via a fax or email, the voter is instructed to always return the paper absentee ballot to their state/jurisdiction via conventional mail.

2.1.3 Voting over the Internet

In 2000, FVAP initiated the Voting Over the Internet (VOI) project to determine if ballots could be reliably and securely cast over the Internet [15, 16]. The project was designed to mimic the established absentee voting process (see section 2.2 for a detailed description of the UOCAVA voting process). Voters who used the VOI system were required to obtain a Department of Defense (DoD) Public Key Infrastructure (PKI) digital certificate used for authentication and web browser plug-in software used to display and transmit ballots to servers administered by FVAP. A voter would use an electronic version of the FPCA to request an absentee ballot and digitally sign the FPCA using the DoD PKI digital certificate. When an electronic absentee ballot request was made, local election officials were notified of the request to be processed. Once a local election official approved the electronic absentee ballot request, a blank electronic ballot was placed on a FVAP server for retrieval. Using a web browser and plug-in, the blank electronic ballot would be retrieved, completed, encrypted, and the encrypted ballot digitally signed by the voter. The encrypted and signed ballots were placed on an FVAP server for retrieval by two local election officials. Note that the completed ballots stored on the FVAP servers were encrypted so that only the local election officials associated with the specific ballot could decrypt the ballots. Once decrypted, the electronic ballots were printed out so that they could be processed (tabulated) in the same way as mail-in absentee ballots. As part of the project, the voters who used the VOI system were allowed to cast traditional paper based ballots.

2.1.4 SERVE

In 2002, the FVAP established the Secure Electronic Registration and Voting Experiment (SERVE) in response to Section 1604 of the National Defense Authorization Act for Fiscal Year 2002. Section 1604 directed the Secretary of Defense to carry out a demonstration project to enable uniformed service members to cast ballots through an electronic voting system by the 2004 general election. SERVE used a web-based architecture with servers hosted and administered by the FVAP. In general, SERVE provided the general capability to electronically identify and authenticate users (voters and local election officials) of the system using unique digital identities (enabled by digital signatures). Voters and local election officials would have to register to become users of SERVE and receive a digital identity. Voters could connect to servers hosted by FVAP to register to vote, request a blank electronic absentee ballot, and complete and return the absentee ballot electronically. Local election officials would connect to servers hosted by FVAP to receive information for voter registration, to receive requests for blank absentee ballots, to distribute electronic blank absentee ballots, to receive completed electronic ballots, and, optionally, ballot tabulation and reports.

In 2003, the FVAP assembled a Security Peer Review Group (SPRG) to review security aspects of the SERVE project. In January 2004, some of the SPRG members released a report highlighting concerns with the security of SERVE [14]. However, no official report was released from the complete SPRG membership. Later in 2004, the Secretary of Defense suspended the

SERVE project. The "Ronald W. Reagan National Defense Authorization Act for Fiscal Year 2005" called for the Secretary of Defense to wait until the EAC established electronic absentee voting guidelines before conducting another electronic voting demonstration project. In addition, HAVA calls for the EAC to consult with the Secretary of Defense to study best practices for facilitating voting by absent uniformed and overseas citizens. It should be noted that UOCAVA remote voting demonstration projects continue to be implemented by state and local election officials as well as public and private organizations. For example, Okaloosa County, Florida is conducting the Okaloosa Distance Balloting Project (ODBP) in partnership with the Operation BRAVO (Bring Remote Access to Voters Overseas) Foundation and the Center for Security and Assurance in Information Technology (C-SAIT) at Florida State University. ODBP placed voting kiosks in three overseas locations that allowed overseas voters to cast ballots in the November 2008 general election.

2.1.5 Interim Voting Assistance System

In September 2004, the Department of Defense launched the Interim Voting Assistance System (IVAS 2004) to allow eligible absentee voters to request and receive absentee ballots over the Internet [16]. To participate in IVAS, users would have to be in the Defense Enrollment Eligibility Reporting System, a US citizen covered by UOCAVA, and already registered to vote in a participating jurisdiction. A voter would connect to the IVAS website running on a FVAP server using Secure Socket Layers (SSL) to request blank absentee ballot. Once a request was made, the appropriate local election official was notified of the request. After the local election official approved the request, the voter was notified via email that their ballot was ready. The voter would connect to the IVAS server via a secure connection in order to download and printout the blank absentee ballot. The voter would use traditional mail to send the completed printed ballot back to the local election official.

In September 2006, the Department of Defense launched the Integrated Voting Alternative Site (IVAS 2006), previously known as the Interim Voting Assistance System (IVAS), to assist UOCAVA voters [17]. IVAS consisted of two tools to request and receive blank absentee ballots– one using purely email messages, the other using a web server running the Secure Socket Layer (SSL) protocol. Both tools required a unique DoD identifier possessed by uniformed service members, their family members, and overseas DoD employees and contractors. The IVAS 2006 identifier requirement limited the UOCAVA population that could use IVAS 2006. Tool One used email messages to allow voters to request blank absentee ballots from their jurisdiction. Using the unique DoD identifier, the voter connected to Tool One over the Internet and logged on to get an electronic version of the Federal Post Card Application (FPCA) form to complete. Once the electronic FPCA was complete, the voter saved the completed electronic form on the local disk of the computer system used to connect to Tool One. The voter attached the completed electronic FCPA form as a PDF file to an email message sent to their local election official. It should be noted that the email sent to the local election official was not electronically/digitally signed by the voter.

The local election official received the blank absentee ballot request email and processed the request. If the absentee ballot request was approved, the local election official provided a blank absentee ballot via fax, email, or traditional mail based on the governing election law. After receiving the ballot, the voter printed out the blank ballot and returned the completed ballot back to the local election official. Tool Two used a secure server to allow the request and delivery of the blank absentee ballots. Using the unique DoD identifier, the voter connected to the protected Tool Two server over the Internet, by the SSL protocol. The voter then completed an electronic version of the FPCA form that was saved to the Tool Two server for processing by a local election official. A local election official then connected to the Tool Two server over an Internet communication protected using the SSL protocol to download the blank absentee ballot request for processing. If the blank absentee ballot request was approved, the local election official posted a PDF file containing the blank absentee ballot. Then the voter securely reconnected to the Tool Two server to retrieve the blank absentee ballot and print the ballot. The voter then completed the blank absentee ballot and returned the completed ballot to the local election official. Neither Tool One nor Tool Two supported the return of completed absentee ballots electronically to the local election officials. Both tools only enabled voters to request and receive blank absentee ballots. It was up to the voter to return the completed ballots back to local election officials using mechanisms outside of IVAS 2006. These mechanisms included fax, e-mail, and traditional mail. In addition, it should be noted that both IVAS 2004 and 2006 did not provide the functionality for a user to register to vote in a jurisdiction. In IVAS 2004 and 2006, the user had to already be a registered voter in a given jurisdiction.

2.2 Current UOCAVA Voting Process

The Department of Defense has implemented several different UOCAVA voting projects (ETS, VOI, SERVE, IVAS 2004, and IVAS 2006) over the last few years. Based on the workflows supported by the DoD UOCAVA projects, several general steps in the UOCAVA voting process can be identified. This section briefly describes the general steps of the UOCAVA voting process.

Step 1: The first general step in the UOCAVA voting process is to have the overseas citizen obtain a voter registration form in order to become a registered voter in the appropriate jurisdiction. Based on a jurisdiction's election laws, a voter could register to vote either before or while the voter is overseas or not in the jurisdiction physically. When a voter registers to vote while overseas, the voter would have to obtain the voter registration form via traditional mail or some electronic means such as fax, email, or website based on the jurisdiction's election laws. Once the voter receives the voter registration form, the voter will complete and return (via fax, email, website, or traditional mail) the form as prescribed by the jurisdiction. If a voter is currently registered to vote in the appropriate jurisdiction, the voter need not complete a voter registration form. The voter registration process for UOCAVA voters is facilitated by the use of the Federal Post Card Application (FPCA) either in paper or electronic forms based on a jurisdiction's election laws to register UOCAVA voters.

Step 2: The second general step in the UOCAVA voting process is for the voter to request a blank absentee ballot from the jurisdiction in which registered. Based on a jurisdiction's election law, a voter could request a blank absentee ballot either before or while the voter is overseas or not in the jurisdiction physically. When a voter requests a blank absentee ballot before going overseas or being physically away from the jurisdiction, a voter may be able to obtain the blank absentee ballot request form physically from a public location (such as the election office, library, office of motor vehicles, etc.), have the form sent via traditional mail, electronically receive the form from a website or email from the jurisdiction, or be required to physically pickup the form from the jurisdiction's election office. If a voter requests a blank absentee ballot while overseas or not in the jurisdiction physically, the voter would have to obtain the blank absentee ballot request form via traditional mail or some electronic means such as fax, email, or website based on the jurisdiction's election laws. Once the voter receives the blank ballot request form, the voter will complete and return (via fax, email, website, or traditional mail) the form as prescribed by the jurisdiction. In addition to facilitating voter registration, the Federal Post Card Application (FPCA) can be used to request a blank absentee ballot either in paper or electronic form based on a jurisdiction's election laws. If a blank absentee ballot cannot be requested by a voter from the jurisdiction in time for a general election, the voter can complete the Federal Write-in Absentee Ballot (FWAB) for Federal offices (such as President/Vice President, U.S. Senator, and U.S. Representative).

Step 3: The third general step in the UOCAVA voting process is for local election officials to process the voter registration forms and blank absentee ballot requests. When a complete voter registration and blank absentee ballot request is received, the local election official will verify the voter's eligibility. If the voter is eligible to vote in the jurisdiction (including voter registration deadline date) and has met the blank absentee ballot request deadline date, the local election official will determine the proper ballot style for the voter and send the blank absentee ballot to the voter via traditional mail or some electronic means such as fax, email, or website based on the jurisdictions election laws.

Step 4: The fourth general step in the UOCAVA voting process is for the voter to receive (via fax, email, website, or traditional mail) the blank absentee ballot from their jurisdiction. When the blank absentee ballot is received, the voter completes the ballot either by printing and marking the ballot physically or electronically completing the ballot with the assistance of a web browser, kiosk, or other application software. Once the absentee ballot is completed, the voter may need to provide additional verification information such as a physical/digital signature or personal identification number (PIN) and date before returning the completed absentee ballot to the jurisdiction. After all jurisdictional requirements are completed, the voter will return the completed absentee ballot to the jurisdiction via traditional mail or some electronic means such as fax, email, or website based on the jurisdiction's election laws. If a blank absentee ballot is not received from the voter's jurisdiction, the voter can complete and return the Federal Write-in Absentee Ballot (FWAB) for Federal offices (such as President/Vice President, U.S. Senator, and U.S. Representative) to their jurisdiction.

Step 5: The fifth general step in the UOCAVA voting process is for the completed absentee ballots, including the Federal Write-in Absentee Ballots (FWABs), to be received for processing by the local election official. Once completed absentee ballots are received via traditional mail or via some electronic means such as fax, email, or website, the local election official will verify that the completed absentee ballots are valid. A local election official will verify that verification information such as physical/digital signatures and/or personal identification number (PIN) are valid, that the ballot was postmarked and/or received by the jurisdiction's deadline dates for absentee ballot return, and that the absentee ballot was completed as required by the jurisdiction (such as limited or no over voted races, use of only pencil or pen to mark choices, etc.). If the absentee ballot verification information (signatures and/or PINs) is valid, the ballot is received before the absentee ballot return deadlines, and the ballot is completed as required by the jurisdiction, the local election official can include the absentee ballot as part of the election's tally based on the jurisdiction's election laws.

2.3 Difficulties in the Current UOCAVA Voting Process

Although there is a general UOCAVA voting process currently used by overseas citizens, there are several difficulties in the process that need to be addressed.

One of the greatest difficulties is the time required to use traditional mail as a mechanism to distribute and receive election material (absentee ballot requests, blank absentee ballots, etc.). In general, the delivery times for postal and military mail to citizens overseas vary greatly depending where the citizen is located. It can take 5 to 10 days for most mail to be delivered to overseas citizens not in the military [25]; and 10 to 14 days for mail to be delivered to military personnel [24]. In addition, uniformed military and overseas citizens may not be at a given physical location for an extended period of time. Given that some jurisdictions finalize their ballots only 30-45 days before an election, using mail to distribute, receive and return election information can be difficult. In some cases the delivery times to distribute blank ballots and return them to local election officials could exceed the window of time between ballot printing and Election Day. This does not take into account the time required for election officials to process and handle blank ballots, or the time required for voters to fill out their ballots and drop them in the mail.

Another difficulty arises when voters use the emergency back-up mechanism for UOCAVA, the Federal Write-In Absentee Ballot (FWAB). First, the FWAB only covers Federal offices (e.g., President/Vice President, U.S. Senator, and U.S. Representative). In general, the FWAB does not allow a voter to vote on state or local questions, although some states will accept write-ins for state-wide offices on FWABs. Since the FWAB is a write-in ballot, the way the voter writes in a candidate's name on the ballot may impact the validity of the ballot based on a jurisdiction's election law. For example, mis-spelling a candidate's name (such as Bil for Bill) or not selecting the official candidate name (such as William, Bill, Billy, Will, Willy, etc.) could impact the ballot validity.

Finally, there are some difficulties common to absentee voting. One such difficulty is with signature verification. Signatures are the most common method for authenticating voters. However, verifying signatures is a difficult task. In order to verify a signature, a trusted sample signature must be on file with election officials. Comparing a received signature with a signature on file requires a great deal of training, although automated signature verification applications may make this task easier.

3 UOCAVA Voting Process

The basic five-step absentee and UOCAVA voting process outlined in Section 2.2 can be simplified and split into three stages: voter registration and ballot request, ballot delivery, and ballot return. This paper identifies and analyzes the use of several options for transmitting election materials for each of these stages. In this section we briefly describe the three stages of the overseas voting process. In each case we identify the types of information exchanged during that stage. The sensitivity of that information, combined with how it will be used during the election, determine the security needs of overseas voting systems implementing each stage, based on the potential impact of a violation of one or more of the security objectives. Federal Information Processing Standard (FIPS) 199, *Standards for Security Categorizations of Federal Information and Information Systems,* [1] identifies and defines these objectives. Table 1, taken from FIPS 199, summarizes these definitions. Later sections of this paper will focus on how various transmission options could support each stage of the overseas voting process, and the threats to these types of systems.

Security Objective	Potential Impact		
	Low	Moderate	High
Confidentiality Preserving authorized restrictions on information access and disclosure, including means for protecting personal privacy and proprietary information. [44 U.S.C., SEC. 3542]	The unauthorized disclosure of information could be expected to have a limited adverse effect on organizational operations, organizational assets, or individuals.	The unauthorized disclosure of information could be expected to have a serious adverse effect on organizational operations, organizational assets, or individuals.	The unauthorized disclosure of information could be expected to have a severe or catastrophic adverse effect on organizational operations, organizational assets, or individuals.
Integrity Guarding against improper information modification or destruction, and includes ensuring information non repudiation and authenticity. [44 U.S.C., SEC. 3542]	The unauthorized modification or destruction of information could be expected to have a limited adverse effect on organizational operations, organizational assets, or individuals.	The unauthorized modification or destruction of information could be expected to have a serious adverse effect on organizational operations, organizational assets, or individuals.	The unauthorized modification or destruction of information could be expected to have a severe or catastrophic adverse effect on organizational operations, organizational assets, or individuals.
Integrity Guarding against improper information modification or destruction, and includes ensuring information non-repudiation and authenticity. [44 U.S.C., SEC. 3542]	The disruption of access to or use of information or an information system could be expected to have a limited adverse effect on organizational operations, organizational assets, or individuals.	The disruption of access to or use of information or an information system could be expected to have a serious adverse effect on organizational operations, organizational assets, or individuals.	The disruption of access to or use of information or an information system could be expected to have a severe or catastrophic adverse effect on organizational operations, organizational assets, or individuals.

Table 1: Potential Impact Definitions for Security Objectives [1]

3.1 Voter Registration and Ballot Request

Description:

Voters register their names and legal voting residences with their local elections officials and request that blank ballots be delivered using postal mail, or some other electronic delivery method. This usually requires that voters provide some form of contact information, such as a mailing address, an e-mail address, or a fax number. The voter provides, or receives, and authenticator which can be used to verify that future correspondence. Typical authenticators include a voter's signature, a Personal Identification Number (PIN), or a digital signature and corresponding certificate.

Information Types:

Voter name, residency information, mailing address
Voter authenticator (e.g. signature, PIN)
Voter identifiers (e.g. social security, driver's license and/or passport numbers)

Security Objectives:

Confidentiality: High
Integrity: Medium
Availability: Medium

Transmission Options:

Postal mail, telephone, fax, e-mail, web-based.

General Issues:

Leaking sensitive personal information from voters.
Available and integrity of voter registration database.

3.2 Ballot Delivery

Description:

Election officials send a physical ballot, or a digital copy of a ballot, to all voters who have requested a ballot. Officials must determine the proper ballot style and send it to the voter using the contact information provided in the ballot request stage. In most cases, outgoing ballots contain tracking information that will be used by election officials when voted ballots are returned.

Information Types:

Candidate and Race information
Possible ballot tracking identifier

Security Objectives:
Confidentiality: Low
Integrity: High
Availability: High

Transmission Options:
Postal mail, fax, e-mail, web-based.

General Issues:
Voters must receive blank ballots in sufficient time to be able to return them to election officials before any deadlines.
Voters must receive the proper ballot styles, determined by their residency information.
Voters must receive blank ballots free from unauthorized modifications.

3.3 Ballot Return

Description:
Voters make their selections on their ballots and return the voted ballot to their local election officials. In nearly all cases, the voter will include an authenticator which can be used to verify the voter's identity. In many cases, the voted ballot includes tracking information that is used by election officials to verify that the returned ballot is the same one that was sent to the voter.

Information Types:
Voter name, address(es)
Voter authenticator (e.g. signature, PIN)
Voter identifiers (e.g. social security, driver's license and/or passport numbers)
Ballot   choices

Security Objectives:
Confidentiality: High
Integrity: High
Availability: High

Transmission Options:
Postal mail, telephone, fax, e-mail, web-based.

General Issues:
Unauthorized individuals returning voted ballots.
Unauthorized individuals modifying voted ballots prior to ballot counting.
Improper disclosure of sensitive personal information from voters or voters' selections.

4 Description of Transmissions Options

The purpose of this report is to identify options for distributing election materials to UOCAVA voters. This section will identify several different transmission options and provide brief descriptions for how these technologies and methods could be used to support overseas voting. The descriptions presented in Sections 4.2, 4.3, and 4.4 are merely examples of typical methods for employing the transmission options. This paper will outline threats to the types of systems described in this section, but other types of systems are possible.

4.1 Transmission Options

This report considers the use of five different transmission options for the distribution and return of election materials: postal mail, telephone, fax, electronic mail, and web-based systems. This section briefly describes each of these transmission options.

4.1.1 Postal Mail

As indicated in Section 2.2, most communication between overseas voters and election officials takes place via United States postal mail, possibly in conjunction with the military postal service. In this case, a voter sends a form via first class mail to his or her local election official's office. Information, such as ballots, is returned by the official to the voter using the address on file, usually from the voter registration phase. The postal service is trusted to reliably transport these materials in a reasonable amount of time, without modifying or reading the contents of the packages. Undeliverable mail, such as when the destination address does not exist, is returned to the sender.
A thorough discussion of the deficiencies in such a system was included in Section 2.3.

4.1.2 Telephone

The Public Switched Telephone Network provides instant two-way communication between nearly any two telephones in the world. The telephone network is a global circuit-switched network consisting of a digital communications backbone with automated telephone exchanges routing calls to their destinations, and, in most cases, with an analog bridge from the backbone to end users' telephones.

Information can be communicated over the telephone network either verbally or by entering numbers on the touch-tone dial pad. In telephone voting systems, voters could communicate authentication information verbally or using the touch-tone dial pad. For instance, voters could enter a PIN on the dial pad, or answer questions verbally in a knowledge-based authentication system. In addition, it may be possible to use Caller ID information to partially authenticate voters.

4.1.3 Fax

Fax machines scan a document and transmit an encoded representation of it over the telephone network to another fax machine. The receiving fax machine can decode the information and print a copy of the scanned document. Some fax machines create an analog representation of the document in a manner similar to analog television, while newer fax machines create a digital representation. The digital or analog representation is sent to the telephone network using analog signals.

Fax machines allow users to transmit written or printed information to another party. In many cases, they are used directly as an alternative to postal mail, allowing voters or election officials to fax election forms or ballots to the other party.

As is the case with telephone communication, telephone network operators are trusted to route faxes to the correct destination based on the number dialed, and not to modify or read faxes in progress.

4.1.4 Electronic Mail

Electronic mail, or e-mail, allows an individual to send text and/or files from one computer to another. This uses the Internet as a communications channel. Thus, the e-mail is transmitted from the sender's computer to his or her mail server (often operated by his or her Internet Service Provider, or ISP), and routed through a series of intermediate servers before being delivered to the recipient's mail server (often operated by an ISP, workplace or a commercial e-mail provider such as Gmail or Yahoo).

In the context of UOCAVA voting, in most cases, information transferred over e-mail would be sent with a form or ballot attached to the e-mail. In some cases it may be necessary for the sender to scan the form or ballot and save it in PDF [8] or other digital format in order to e-mail it.

Using standard e-mail, the recipient of a message does not receive any assurance of the identity of the sender, as it is easy to forge a return e-mail address. The sender may receive some assurance that the recipient received the e-mail. Many e-mail servers will send a warning to the senders of undeliverable e-mail. However, some e-mail servers, in order to limit unsolicited e-mails, do not sending these warnings.

E-mail can be encrypted. The current standard for e-mail encryption using Public Key Cryptography is the Secure/Multipurpose Internet Mail Extensions (S/MIME) protocol [22]. Most major e-mail clients include S/MIME functionality; however use of S/MIME encrypted e-mail is relatively rare. Use of S/MIME requires all users to have a public/private key pair and be part of a Public Key Infrastructure. Furthermore, commonly used web-based e-mail providers do not include S/MIME functionality. Because of the limited deployment and usage of S/MIME, this paper will assume e-mail communications are unencrypted unless otherwise noted.

4.1.5 Web-Based

It is also possible to use web sites to communicate between two parties. While both web sites and e-mail use the same communication channel, the Internet, the two options use different communication protocols. Also, the user experience in the case of a web site is vastly different than that of e-mail. The interface can be customized, and the overall experience is more interactive.

A web-based UOCAVA voting system would include a web server operated by a local election official. That official could post information for all to see, such as blank registration forms, or blank ballots for each precinct. If this material is posted as a document, users could download files, print them, and return them to the official using some other form of communication. If the materials are posted as web forms, users could fill in the information on the web site and return it, in a manner similar to filling out billing information after purchasing something online.

Alternatively, the web site may grant different users access to different information. For instance, upon registration each voter would be given a username and password for the site. Upon logging on to the site, the voter would only have access to relevant information for him or her; for example, the voter would only see his or her ballot.

Properly developed and configured web sites can contain additional security protections not found in e-mail by using SSL (Secure Socket Layers) or TLS (Transport Layer Security) [4,7]. This would allow for encrypted communications between the web server and a voter to prevent eavesdropping. Digital certificates could be used to give voters assurance they are on the correct website. A more detailed discussion of security controls is presented below.

4.2 Options for Voter Registration and Ballot Request

The previous section discussed five different transmission options for voting materials. The next three sections discuss how each of these options could be used to support the three stages of overseas voting. As previously noted, this section outlines typical election systems using the transmission options, but does not attempt to capture every possible variation.

The first stage of the UOCAVA voting process is the registration and ballot request stage. In this stage voters submit registration information confirming their identities and places of residence, and provide election officials with contact information. This section describes how election materials from this stage could be sent using postal mail, telephones, fax machines, electronic mail, and web-based systems.

4.2.1 Postal Mail

As discussed in Section 2.2, all states accept the Federal Post Card Application (FPCA) to register military and civilian overseas citizens to vote and for requesting ballots. Voters obtain these forms from a variety of locations, including military voting assistance officers, embassies and consulates. Some web sites, such as the Overseas Vote Foundation [27], have posted copies of the form. Voters unable to find an FPCA may request one from military service departments or the State Department.

The FPCA asks each voter for his or her name, voting residence address, mailing address and additional contact information. This information is used to determine voter eligibility, contact voters if problems are discovered, and distribute voting materials, such as absentee ballots.

The FPCA is also used to establish a shared authenticator that election officials can use to verify future correspondence from the voter, in this case the voter's signature. To gain some level of assurance that the person who filled out the form is the individual claimed, the FPCA asks for the voter's military identification number or passport number. If a voter is unable to provide either of those, some states require a notary to sign the FPCA.

4.2.2 Telephone

The public telephone network could be used to exchange voter registration information. In this case voters could obtain the telephone number for their local election official and call to register to vote or request a ballot. Voters would speak to either an election official or an automated registration system, providing their name, voting residence address, and any contact information required, such as a telephone number or mailing address.

In order to authenticate the registration, each voter would need to provide sensitive, identifying information, such as a military identification number or passport number, which election officials could verify. Voters unable to provide the required identifying information would not be able to register over the phone. The election official and voter may use this time to establish a new shared authenticator for future correspondence, such as a PIN or a password. Alternatively, election officials and voters may continue to use the identifying information used to verify the voters' identities.

4.2.3 Fax

Several states allow voters to fax completed FPCAs to their local election officials. The procedures for marking and returning FPCAs are the same as for postal mail (see Section 4.2.1), except that the completed form is faxed to the local election official rather than mailed. The election official should have a dedicated fax line for receiving FPCAs, and this machine should be kept in a secure room.

4.2.4 Electronic Mail

Some states allow voters to e-mail completed FPCAs to their local election officials. In this case, each voter would have to obtain a paper copy of the FPCA, either by finding a physical copy of the form or printing an electronic version. The voter would sign the paper FPCA, and use a scanner to save it on his or her personal computer in a standard file format, such as the Portable Document Format (PDF). The resulting file could be sent as an attachment in an e-mail to a special e-mail address set up by election officials for registration and ballot requests.

In the typical case described above, a voter's signature is required in order to authenticate the source of the registration form. Election officials may be able to compare the signature on the form to voter registration information on file. Individual jurisdictions may determine that other information could be used to authenticate the voter's identity. This could include requesting confidential personally identifiable information that is verifiable by election officials. Digital signatures would provide an alternative method for authenticating voters. Voters with a public/private key pair could digitally sign their registration forms, which could be verified by election officials upon receipt. Digital signatures would be nearly impossible to forge, and the process would not put sensitive personal information at risk of being intercepted. However, it would require a large-scale Public Key Infrastructure, which does not yet exist.

4.2.5 Web-Based

Voters could submit registration and ballot request information on an election official-operated web site. Voters could fill in registration information directly on the web site from an Internet browser, and submit the information without printing or scanning any forms. Web servers could implement cryptographic protocols (e.g. SSL/TLS) to protect information as it is transmitted to and from the voters.

Such a system could not rely on voter signatures for authentication purposes. Web-based registration would have to rely on other methods for voter authentication, such as those described in Section 4.2.4.

4.3 Options for Ballot Delivery

The second stage of the UOCAVA voting process is the delivery of the ballots. In this stage, election officials send blank ballots to voters using the contact information submitted during the registration and ballot request phase. This section describes how blank ballots could be sent using postal mail, fax machines, electronic mail, and web-based systems. Telephone systems are not considered in this section, as any telephone voting system would also incorporate a mechanism for making ballot selections. Telephone voting systems will be discussed in the next section.

4.3.1 Postal Mail

Election officials begin to distribute paper ballots after they are printed. Upon receiving a ballot request from a voter, election officials look up the voter registration status of the voter and, once confirmed, determine the proper ballot style for that voter's precinct. The ballot is then sent to the mailing address indicated by the voter's ballot request. The complete package usually contains instructions, return envelopes and other items to facilitate the ballot marking and return process. These items will be discussed when postal mail ballot return is discussed.

To track the ballot request through the delivery process, officials indicate in their records that a particular ballot request has been accepted, processed and sent out. In some cases, identifying information is passed along with the ballot during the processing and delivery of the ballot. It is important to note that this information is not printed on the ballot, but rather it is a physically separate item that follows the ballot. For instance, it could be a barcode printed on the outside of an envelope containing the ballot.

4.3.2 Telephone

Ballot delivery via the public telephone network would only work in the context of a vote by phone system. This option will be discussed in the next section.

4.3.3 Fax

Blank paper ballots could be faxed to voters as an alternative to postal mail. Most of the process is similar to postal delivery of ballots. Upon receiving a ballot request from a voter, election officials look up the voter registration status of the voter and, once confirmed, determine the proper ballot style for that voter's precinct. Again, this ballot does not have any identifying marks that could tie a particular ballot back to a particular ballot request or voter. The ballot, along with ballot marking and return instructions, is faxed to the number listed on the voter's ballot request.

Detailed ballot tracking procedures are not necessarily required for delivery of blank ballots via fax. Election officials receive immediate notification that the ballot was successfully delivered to the voter's requested fax machine. However, tracking numbers may be used internally by election officials prior to faxing the ballot in order to track the ballot request and delivery process at the election offices. These numbers may also be used to identify that the same ballot faxed to a particular voter is the one returned by that voter.

4.3.4 Electronic Mail

As in the fax and postal mail options, upon receiving a ballot request, officials check the registration status of the voter and determine the appropriate ballot style. As in the processes described previously, this ballot should not contain any identifying marks that could be tied back to a particular voter. In this case, the ballot must be in a digital form, such as in a Portable Document Format (PDF) file [8]. Officials could have digital copies of all ballot forms, or they could construct digital ballots from paper ballots using a scanner.

The ballot is sent as an attachment from an election office computer in an e-mail to the voter-provided e-mail address. Marking and return instructions should accompany the ballot, usually as plain text in the e-mail message. As with any e-mail message, the message travels from the election office computer, to the office's Simple Mail Transfer Protocol (SMTP) server [9]. From there the server determines how to route the message to the recipients e-mail address. In most cases the message will pass through a series of intermediate network devices before arriving at the recipients e-mail server. The message will remain on the server until the recipient logs into their e-mail account. Depending on the e-mail protocol used by the recipient the message may be deleted off the server after being accessed by the voter. Generally, webmail providers retain copies of e-mails. Other providers, such as internet service providers, often provide POP3 service, which allows voters to download copies of e-mails, which are then promptly deleted from the server.

As previously mentioned, most e-mail servers will send error messages to the e-mail sender if the message is not deliverable (for instance, if the address does not exist, or if a server is malfunctioning). Therefore, election officials should, at a minimum, follow up on all returned e-mail messages with other forms of communication. For additional protection against undeliverable mail, officials could request return receipts from recipients. Such receipts are automatically generated by recipient computers and delivered to the sender when an e-mail message is actually read by the voter, as opposed to simply being delivered to the voter's e-mail server.

4.3.5 Web-Based

Rather than sending digitized ballots to voters individually, jurisdictions could post ballots on a public web site and instruct voters to obtain their ballots via that site. When discussing web-based delivery of ballots in this paper, we will assume that ballots will be returned via postal mail, fax or electronic mail. Thus the posted ballots would be in a digital format, such as PDF, suitable for printing. We discuss web-based delivery and return of ballots in the next section.

For the purposes of this paper, we consider a web-based ballot distribution system that is connected to the voter registration database. After registering to vote via some other method, voters could navigate to the election web site. The site would prompt each voter for identifying information, such as his or her name, date of birth and a portion of his or her street address. This information is not used to strongly authenticate the identity of the voter, but rather to look up the voter in the registration database to determine the proper ballot style and present it to the voter. After downloading the ballot, the voter would mark the ballot on the computer or print it and mark it by hand. Ballots would be returned using postal mail, fax or electronic mail.

4.4 Options for Ballot Return

The third stage of the UOCAVA voting process is ballot delivery stage. In this stage voters return voted ballots to their local election officials. This section describes how voted ballots could be sent using postal mail, telephones, fax machines, electronic mail, and web-based systems.

4.4.1 Postal Mail

After receiving a physical or electronic blank ballot, a voter may, if necessary, print a paper ballot, and then make his or her selections on the ballot. In most jurisdictions, the voter is instructed to place the ballot in a privacy envelope, which may be a standard envelope or one provided by election officials. The privacy envelope is placed in an outer envelope, along with information used to authenticate the voter and the voted ballot (or this information is written on the outer envelope), creating a single package of voting material. This envelope may be placed in an additional return envelope, or placed directly in the mail. Upon delivery, outer envelopes are stored in a secure location until the election polls close and ballot tallying begins.

Multiple envelopes are used to protect voter privacy during the tabulation phase. Election officials open the outer envelope and separate identifying information from the privacy envelope prior to opening the privacy envelope and tallying the votes.

Many jurisdictions use ballot tracking procedures to follow individual ballots throughout the delivery, return and counting processes. Identification numbers and code, often in the form of barcodes, are included on individual ballots, privacy envelopes, outer envelopes, return envelopes, or some combination of those items. This provides some assurance that ballots are not lost during the tabulation process. Furthermore, election officials could use the information on the barcodes to verify that the same ballot that was sent to an individual voter was the one that was returned by that voter, offering some protection against attacks. However, ballot tracking information could be used to violate voter privacy. In many cases, a large portion of the ballot tracking process is performed using automated systems or en masse, which provides some protection against malicious individuals attempting to use tracking information to determine how individuals voted.

4.4.2 Telephone

Telephone voting systems do not have distinct ballot distribution and return stages. Voters are provided with ballot questions and immediately given an opportunity to make selections. Voters would not have to wait for ballot materials to be distributed, but they would have to wait until they have received voting credentials and until the polls open on the telephone voting system.

In most cases, the telephone voting system would be a computer system with connections to several telephone lines. The computer system would automatically receive calls, provide voting instructions, authenticate voters and store cast ballots. Prior to opening the telephone polls, election officials would have to initialize the voting system with information about registered voters, authentication information, and ballot styles for all jurisdictions under their control. Voter information could be initialized using information from the registration and ballot request stage. For example, upon receiving a registration and ballot request, election officials would enter the voter's name and residency information in the voting system. This information would be used to identify the appropriate ballot style for a given voter. Election officials would also generate a random personal identification number (PIN) for the voter, and provide it to the voter and the voting system. The PIN would be used to authenticate the voter.

After the polls have been opened, voters could call the telephone voting system from their personal telephones, supply their name, residency information and PIN for authentication purpose, and cast a ballot by following the prompts on the phone.

Telephone voting systems are currently in use in the state of Vermont. However, the Vermont system is not used for remote voting, but rather to serve as an accessible voting station for visually impaired voters. Voters must still go to their local polling places to vote even if they will use the telephone voting system.

4.4.3 Fax

Fax machines could be used to transmit voted ballots to election officials. After receiving physical or electronic ballots, voters could make their selections on their ballots and print out paper copies, if necessary. Voters may also need to obtain one or more election forms, if they were not delivered via postal mail. These forms would have fields for the voter's name, residency information, signature, and other information needed by the election officials. Additionally, voters may be instructed to sign a form that includes information about privacy issues when using a fax machine to return a ballot. This package of materials, the voted ballot and accompanying forms, could then be faxed to an election official.

Upon receiving the faxed ballot and voter information, an election official would package this information together and store it in a secure location until the tabulation process begins. Unlike postal mail voting, there are no physical protections for maintaining vote secrecy. As part of the tabulation process, election officials would authenticate voters by comparing the voter's signature on the form with the signature on file from the registration process. In some cases, the selections on the faxed ballot are transferred to another ballot, such as an optical scan ballot.

4.4.4 Electronic Mail

Given the wide usage of e-mail in everyday communications, e-mail may be an attractive option for quickly returning electronic ballots to officials. In this paper, we consider a ballot return method using e-mail which closely follows the fax method. This method is already used by several states in the country.

The voting process would be very similar to the process described in Section 4.4.3 for ballots returned via fax. Voters would obtain and mark a paper ballot, and fill out accompanying voter forms for identification purposes. However, rather than faxing these materials to election officials, the voter would scan them on a computer, creating a digital copy of the ballot package, or use some other device capable of scanning and e-mailing attachments. Voters would have to save the scanned materials in a standard file format, such as PDF. The resulting file, or files, could be sent to election officials as attachments in an e-mail.

Upon receiving the ballot package, an election official would open the attachment and print a paper record of the ballot and accompanying voter forms. This package would be stored in a secure location, along with other paper ballots received via fax or postal mail. As was the case with fax return of ballots, there are limited procedural protections that could maintain voter privacy. Election officials charged with responding to e-mailed ballots would have access to voters' identities and ballot selections.

It may be possible to automate additional steps in this process using a computer. Depending on the format of the received ballots, a computer may be able to automatically tally votes as they are received via e-mail. They could also be used to assist election officials in authenticating received ballots. Some absentee ballot management systems even include signature verification functionality. In general, however, such systems are not considered in the threat analysis outlined in this paper.

4.4.5 Web-Based

In this paper, we consider web-based Internet delivery of ballots to be what many refer to as Internet voting. That is, web-based voting is a voting system in which voters make ballot selections and cast their votes on a web site operated by election officials. Like the telephone voting option described in Section 4.4.2, web-based Internet voting does not require a separate ballot delivery stage. Note that this paper considers web-based ballot delivery and web-based ballot return as two different types of voting systems. Section 4.3.5 covers only the distribution of blank ballots, and assumes some other method will be used to return voted ballots to election officials. This section assumes the web site will allow voters to both view ballot contests and cast ballots with their selections.

Web-based Internet voting systems consist of an election web server connected to the Internet. The server would have similar functionality to the telephone system described in Section 4.4.2, in that it would authenticate voters, provide ballot contests, and record voters' selections. Voters would connect to the election web server from computers using a standard web browser.

Prior to opening the polls, election officials would have to initialize the voting system with information about registered voters, authentication information, and ballot styles for all jurisdictions under their control. For example, upon receiving a registration and ballot request, election officials would enter the voter's name and residency information in the voting system. This information would be used to identify the appropriate ballot style for a given voter.

The voting system would rely on the voter authenticator exchanged during the voter registration and ballot request stage. More traditional methods for absentee voting rely on voter signature verification for authentication purposes, which would not be possible in a web-based voting system. Typical authentication methods for web-based Internet voting include digital signatures, PINs and passwords. NIST SP 800-63, Electronic Authentication Guideline, [5] discusses several methods for remote authentication which could be used in an Internet voting system.

5 Threat Analysis Methodology

The remainder of this paper focuses on the security issues related to using these types of systems. Section 5 contains a threat analysis for each of the 14 systems considered in Section 4. This analysis was performed based on methodology provided in NIST SP 800-30, *Risk Management Guide for Information Technology Systems* [2], with some important modifications. The first step in the threat analysis is characterizing the election systems. Typically this is done with a particular system in mind, knowing what type of information will be handled, what procedures will be followed, and what equipment will be used. This report, however, looks at systems from a high level, where none of these items is known with any amount of specificity. The high level descriptions of transmission options for each stage of the voting process given in Section 4 characterize the systems analyzed in this report. As these characterizations are high level, the threat analysis must be performed at a correspondingly high level.

For each system, we identified methods (i.e., threats) for attackers to violate one of the major security goals of the election system: confidentiality, integrity and availability. We then consider the level of access to election systems, skills and resources that would be needed to carry out a threat. Based on that analysis, we identify a set of groups or individuals capable of carrying out a threat, and estimate the likelihood that election officials would be able to detect an attack from that group or individual. Finally, we propose security controls that could mitigate or eliminate the identified threat. The following subsections describe each of these stages in more detail.

5.1 Threats

Threats are events or circumstances that are potential violations of security. For each transmission option we list high-level threats that describe potential security problems. For example, a threat could involve compromising the privacy of votes, modifying cast ballots or making the voting system inaccessible to voters. Not all threats are caused by humans; natural disasters and equipment failures are potential threats, particularly to the availability of systems. However, this report focuses on threats, such as those from malicious individuals or groups, as these threats can attack any of the security objectives of a system in a variety of ways.

5.2 Threat Sources

Threat sources are groups or individuals that could feasibly attack a voting system. Some attacks on voting systems could be conducted by almost any dedicated individual, while others may require significant resources, knowledge or access to voting system equipment. Threat sources can be broken down into two classes: internal and external sources. Internal sources are individuals or groups with some level of authorized access to the voting system equipment or the supporting infrastructure (e.g. the communications network). External sources are individuals or groups that do not have any special level of authorized access to the voting system equipment or supporting infrastructure. This report considers the following examples of threat sources.

Internal Threat Sources:
- **Legitimate Voters**: Legitimate voters have a limited level of access to voting system equipment. That is, each voter is allowed to submit registration information, obtain the proper ballot given their registration status, and cast a single ballot. Voters may, for example, attempt to use or expand their authorized level of access to damage the election system, change the results of the election, or harm the credibility of the election results.

- **Election Officials**: Election officials have a significant level of access to data on voting system equipment. They are users of the election system with access to voter and ballot information, but may not be authorized system administrators. However, while election officials may be restricted from certain administrative functions, such as software installation, they often have relatively unrestricted physical access to voting system equipment. Malicious election officials could use their privileged access to voting systems to exploit the system.

- **System Operators**: While election officials are users of an election system, system operators serve as administrators, ensuring that the systems function properly or seeing that vital operations are fulfilled. System operators may administer the election system directly, or they may administer the supporting infrastructure for the election. For example, postal mail employees, including mail carriers and sorters, would be system operators in elections which use the postal mail as a communications medium. Network technicians at major telephone companies or Internet Service Providers (ISPs) would be examples of system operators when the telephone network or the Internet is used. In all cases system operators have a privileged level of access to equipment that is vital to conducting the election.

- **Other insiders**: Other individuals or organizations may have privileged access to voting system equipment, either before, during or after an election is conducted. For example:
    - Voting System Manufacturers
    - Voting System Integrators
    - Support staff

External Threat Sources
- **Hostile Individuals**: Individuals without special access privileges to the voting system may attempt to exploit vulnerabilities. In many cases, these individuals would be limited only by their technical knowledge and their ability to deceive individuals with privileged access to the voting system (e.g. social engineering). However, some types of attacks may require multiple attackers acting in unison or significant resources that one person cannot easily accumulate or control.

- **Hostile Organizations**: A hostile organization and a hostile individual differ in the amount of human and technical resources under their control. Hostile organizations would be able to recruit, hire, and train several individuals to participate in an attack. An organization would likely have more resources, both monetary and technical (e.g. computers, network bandwidth). Hostile organizations could take many forms. While their attacks motives may differ, the possible desired outcomes for attacks are likely the

same: controlling the result of the election, disrupting the voting process, or damaging the credibility of the election. Examples of hostile organizations include:
- *Hostile Civilian Organizations*
- *Foreign-Sponsored Organizations*
- *Terrorist Organizations*

5.3 Effort
Effort refers to the relative level of difficulty of performing a successful attack based on a threat. Each threat is classified into one of three levels:
- **Low**: An attack would require little or no resources or detailed knowledge of the system. *Example: Forcing a voter to vote a particular way in the presence of an attacker.*
- **Moderate**: An attack would require significant resources (or an ability to obtain such resources) or knowledge of the system. Inside attacks involving a small number of co-conspirators fall in this category. *Example: A Denial of Service (DoS) attack against election official computers and servers.*
- **High**: An attack would require extraordinary resources, knowledge of the system or access to the system. Inside attacks involving a large number of co-conspirators fall in this category. *Example: Replacing absentee ballots with forgeries during manual hand-counts.*

5.4 Detection
Organizations can recover from or mitigate attacks if they are detected. For each threat, this report estimates the relative level of difficultly of detecting whether a particular threat has been realized in an attack. In general, attacks are more severe when they go undetected. The threat matrix estimates the likelihood that an attack would be detected, and classifies it according to three levels:
- **High**: An attack would most likely be detected given proper monitoring. *Example: An attacker luring voters to an imposter election web site.*
- **Moderate**: An attack may be detectable, but could require a large amount of resources and time. Such attacks are unlikely to be detected during the election. *Example: A computer virus infecting personal computers.*
- **Low**: An attack is unlikely to be detected without extraordinary resources. *Example: Malicious code installed on election equipment by election insiders.*

5.5 Impact
The impact of an attack is its effect on violating the system's basic security objectives. The threat analysis includes *low*, *moderate* and *high* modifiers for each impact. The modifier indicates the likely severity of an attack from a given threat. Severe attacks must impact a significant number of votes or voters, or seriously damage the credibility of the election process. Descriptions of the security objectives and impact levels are described in Section 3, Table 1. These goals are:
- **Confidentiality**
- **Integrity**
- **Availability**

5.6 Possible Controls

Where possible, each threat is accompanied by possible mitigation techniques in the form of security controls from NIST SP 800-53 [3]. These controls are identified by the security control number. Section 7 of this report will discuss these controls in greater detail. In some cases, the systems targeted by an attack are outside the control of election officials. For instance, voters' personal computers are not administered by election officials, preventing officials from protecting those systems. Most threats to systems outside the control of officials do not have any suggested security controls.

6 Threat Analysis

The purpose of this report is to consider various technologies which could be used to improve the UOCAVA voting process and to identify high-level threats associated with each system. This section documents the threats identified using the methodology identified in Section 5. The threat analysis methodology used is a variation of the one outlined in NIST SP 800-30, *Risk Management Guide for Information Technology Systems* [2]. In particular, this report performs a threat analysis on each of the voting system transmission options identified in Section 4 for the three voting stages, Registration and Ballot Request, Ballot Delivery, and Ballot Return. Sections 4.2, 4.3, and 4.4 characterize how this report assumes each of these transmission options will be used in an election. In practice, many jurisdictions may use different procedures and technical controls while conducting elections. Specific threats and threat sources may differ slightly depending on the exact nature of how a particular transmission option is used.

Tables summarize the threats to each transmission option considered for the three stages. The first column of this table identifies the threat (see Section 5.1), while the second column identifies the individuals or groups capable of exercising that threat (see Section 5.2). The next three columns identify the level of effort required to exercise the threat (see Section 5.3), the relative probability that election officials would detect an attack (see Section 5.4), and the impact of the attack succeeding on the election (see Section 5.5). The final column identifies security controls that could mitigate the threat. Security controls are discussed in greater detail in Section 7 of this paper.

6.1 Registration and Ballot Request

This section documents threats to the transmission options for the Registration and Ballot Request stage, as described in Section 4.2.

6.1.1 Postal Mail

The most widely used method for returning registration materials and requesting ballots is via postal or military mail. In this stage, voters send sensitive personal information to election officials to both identify themselves and to establish an address to send future correspondence, such as the blank ballot. One of the major concerns is that attackers could inject themselves in the communications path between the voter and the election official in order to collect personal information. The attacker could use this information to impersonate the voter, or possibly inflict financial damage on the voter (e.g. identity theft) depending on the type of information contained on the registration card.

Items in the mail are handled by a large number of people. In theory, any of the individuals charged with delivering a registration/request form could open the envelope to obtain personal information. However, this threat is substantially reduced by a variety of factors. Most postal carriers undergo some form of background check. Furthermore, it would be extremely difficult for a small number of malicious individuals to obtain a large amount of information. Most postal employees would only handle a small number of registration/request materials. In some cases it might be difficult to identify these materials from other pieces of mail without opening the

Threat	Threat-Sources	Effort	Detection	Impact	Possible Controls
Ineligible individual allowed to register to vote.	Hostile Individuals	Low	Mod.	Integrity-Mod.	IA-1, IA-2, IA-4, IA-5
Valid voter's ballot request information, such as address, is modified in transit.	Hostile Individuals Postal Workers System Operators Election Officials	Mod.	Low	Integrity-Mod. MP-5,	MP-5(1)
Registration/Request materials are accidentally lost or destroyed in transit.	Postal Workers	Low		High Avail.-High	MP-5
Registration/Request materials are intentionally delayed or destroyed in transit by a malicious party.	Hostile individuals Hostile Organizations Postal workers	High		High Avail.-High	MP-5
Sensitive personal information is viewed in transit.	Postal Workers	High		Low Confid.-Mod.	MP-5
Sensitive personal information is improperly read after delivery.	Election Officials	Mod.	Mod.	Confid.-Mod	MP-1, MP-2, MP-4, PE-2, PE-3, PS-2, PS-3
Sensitive personal information is improperly modified after delivery.	Election Officials	Mod.	Mod.	Integrity-Mod.	MP-1, MP-2, MP-4, PE-2, PE-3, PS-2, PS-3

Table 2: Threat Matrix for Postal Mail Registration and Ballot Request

envelopes. Due to these factors, it is unlikely that a large scale loss of personal information could occur during transmission through the postal service.

One of the primary disadvantages of postal and military mail is the transmission time. Registration materials could be lost, destroyed, delayed or intercepted during transit from the voter to the election official. However, delays during registration and ballot request are not as damaging as at other points in the UOCAVA voting process. This stage of the process can occur well before an election, mitigating the damage caused by delays. With adequate lead time before an election, voters could also detect lost or destroyed registration materials, after noticing the absence of a response from election officials after mailing a form.

6.1.2 Telephone

Telephones could be used to transmit registration and ballot request information. One of the major functional differences compared to postal mail systems is that the standard form of authentication information, the voter's signature, would not be available for use. Voter authentication would have to be done using secret, and potentially sensitive, information identifying the voter. Depending on the type of information used, it may be easier for a group or individual to fraudulently register or request ballots for legitimate voters, as compared to processes that use both secret information and voter signatures.

As in the case with postal communication, there is a danger that transmitted personal information could be intercepted by malicious third parties. Information traveling over telephone lines could theoretically be intercepted by anyone with access to the telephone operator's equipment or physical lines. Many people, primarily telephone network employees, would have access to the equipment or lines. However, it would be extremely difficult for an individual or a group to successfully intercept personal information. The most likely scenario would be for an attacker to infiltrate the local central office near the election systems. Sabotaging the telephone network equipment, or jamming the telephone lines, would require a comparable amount of access to

Threat	Threat-Sources	Effort	Detection	Impact	Possible Controls
Ineligible individual allowed to register to vote.	Hostile Individuals	Low.	Mod.	Integrity-Mod.	IA-1, IA-2, IA-4, IA-5, IA-7
Election official offices have too few telephone lines to handle demand.	Telephone Operators System Operators	Low	High	Avail.-High	IR-4, IR-5
A denial of service attack, or other technical attack, jams telephone lines.	Telephone Operators Hostile Organizations	Mod.	High	Avail.-High Integrity-Mod.	IR-4, IR-5, CP-7, CP-8 SC-5, SC-8
Personal information is intercepted between the voter and election official.	Telephone Operators Hostile Organizations	High	Low	Confid.-Mod.	PE-4, SC-8, SC-9, SC-12, SC-13
Disgruntled election official fails to properly record registration information.	Election Official	Mod.	Low	Integrity-Mod.	PS-2, PS-3

Table 3: Threat Matrix for Telephone Registration and Ballot Request

network equipment, but would be significantly easier to conduct. Such an attack would prevent legitimate voters from sending their registration and ballot request information.

A recent development in the area of telephone communications is the adoption of voice-over-internet-protocol (VoIP) technology. Telephones using VoIP use the Internet to transmit calls, rather than the traditional telephone network, the Public Switched Telephone Network (PSTN). There are more opportunities for attackers to eavesdrop, disrupt and modify information on the Internet than the PSTN, particularly if individuals are using wireless access points to distribute their own Internet connection to VoIP devices.

Denial of service attacks are also a major concern. Individual jurisdictions would have a limited number of telephone lines available to them, and perhaps a more limited number of employees staffing them. An organization with significant resources could purchase enough telephone lines to prevent legitimate voters from speaking to election officials.

6.1.3 Fax

Fax machines would be able to transmit both secret information from the voter and the voter signature for authentication purposes. Fax machines use the telephone network to transmit information, so the same concerns about intercepted communications exist for registration via fax as for telephone calls. As previously noted, such attacks would be very difficult to carry out and require access to the telephone network infrastructure.

While a telephone call might be answered by an election official directly, or via an automated electronic process on a computer, fax machines would likely be left in a room at the election office receiving faxes throughout the day. In most cases the machines would be unattended. Received registration and ballot request forms could sit in the fax machine tray for several hours before being processed by an election official. This gives would-be attackers time to view sensitive personal information or destroy valid registration forms.

Threat	Threat-Sources	Effort	Detection	Impact	Possible Controls
Ineligible individual allowed to register to vote.	Hostile Individuals	Low.	Mod.	Integrity-Mod.	IA-1, IA-2, IA-4, IA-5, IA-7
Election official offices have too few fax machines and/or telephone lines to handle demand.	Telephone Operators System Operators	Low	High	Avail.-High IR-4,	IR-5
A denial of service attack, or other technical attack, jams telephone lines.	Telephone operators Hostile Organizations	Mod.	High	Avail.-High Integrity-Mod.	IR-4, IR-5, CP-7, CP-8, SC-5, SC-8
Personal information is intercepted between the voter and election official.	Telephone Operators Hostile Organizations	High	Low	Confid.-Mod.	PE-4, SC-8, SC-9, SC-12, SC-13
Disgruntled election official fails to properly handle faxed registration forms upon receipt.	Election Official	Mod.	Low	Integrity-Mod. PS-2,	PS-3
Sensitive personal information is improperly read from faxed registration forms prior to processing.	Election Officials Support Staff Hostile Individuals	Mod.	Mod.	Confid.-Mod	PE-2, PE-3, PE-5, PE-6, PS-2, PS-3
Sensitive personal information is improperly read from processed registration forms in storage.	Election Officials	Mod.	Mod.	Confid.-High	MP-1, MP-2, MP-4, PE-2, PE-3, PE-6, PS-2, PS-3

Table 4: Threat Matrix for Fax Registration and Ballot Request

Fax machines would not necessarily give voters instant notification that their registration and ballot request forms were received properly. Certain errors on the election official's fax machines, such as low ink, would not be reported back to the voter automatically. Also, voters would not receive automatic notification if they filled out the forms incorrectly.

6.1.4 Electronic Mail

Electronic mail uses the Internet and a computer to transmit information. Voter authentication could be performed with some combination of secret personal information from the voter and a voter signature (the latter would require voters to print, sign and scan a physical paper ballot). There is potential for this information to be intercepted, and possibly modified, en route from the voter to the election official. E-mails travel through telecommunications lines, network equipment and e-mail servers before reaching the intended recipient. As e-mails travel unencrypted throughout the network, anyone with access to the infrastructure could read or even modify e-mail messages. In particular, e-mail servers often store messages for a short period of time before passing them on to the next server, or the intended recipient. System operators for these servers could possibly intercept registration e-mails.

E-mail does not provide any guarantee that the intended recipient will receive the message. The e-mail system relies on the Domain Name System (DNS) to route e-mails to the proper servers. An attack on DNS servers could route e-mails to an attacking party. This would not only result in voter disenfranchisement, but also the loss of sensitive voter information. This kind of attack would require very sophisticated attackers focusing their efforts on major e-mail service providers. There are no known reports of a similar attack being successfully conducted on e-mail or DNS servers. However, it is important to note that a recent vulnerability was discovered in DNS servers that could have been used to construct a similar attack [13]. DNS servers were

Threat	Threat-Sources	Effort	Detection	Impact	Possible Controls
Ineligible individual allowed to register to vote.	Hostile Individuals	Low.	Mod.	Integrity-Mod.	IA-1, IA-2, IA-4, IA-5, IA-7
Voter information from registration/request materials is read or modified on the e-mail servers of the voter or election official by authorized system administrators.	Network Operators	Low	Low	Integrity-Mod.	AC-2, AC-3, AC-5, AC-6, SC-9, SC-12, SC-13
Voter information from registration/request materials is read or modified on the e-mail servers of the voter or election official by unauthorized individuals.	Hostile Individuals Hostile Organizations	High	Mod.	Confid.-High Integrity-High	AC-2, AC-3, AC-5, AC-6, AC-12, SC-9, SC-12, SC-13
A denial of service attack against voter and/or election official e-mail servers overwhelms resources and prevents the transmission of registration/request materials.	Hostile Organizations	Low	High	Avail-High	IR-4, IR-5, CP-7, CP-8, SC-5, SC-7
Election official offices have too few resources (e.g. bandwidth, servers) to handle legitimate traffic.	Network Operators Election Officials	Low	High	Avail-High IR-4,	IR-5
Personal information is intercepted between the voter and election official on the Internet.	Hostile Organizations Network Operators	High	Low	Confid.-High	PE-4, SC-9, SC-12, SC-13
Malicious code (e.g. spyware) on the voter's computer transmits personal information from the registration/request materials to a third party.	Hostile Individual Hostile Organization	High	Mod.	Confid.-High	*Outside control of officials.*
Malicious code (e.g. a Trojan horse) on a voter's computer modifies or disrupts outgoing e-mails for with registration/request information.	Hostile Individual Hostile Organization	High	Mod.	Integrity-High	*Outside control of officials.*
Disgruntled election official fails to properly respond to e-mailed requests.	Election Official	Mod.	Low	Integrity-Mod.	PS-2, PS-3
Voters send registration/request materials to an incorrect e-mail address, resulting in the disenfranchisement and the loss of personal information.	Hostile Individual Hostile Organization	Low	High	Confid.-High Avail.-Mod.	*Largely outside control of officials.*
An attack on the DNS system causes e-mails containing personal information to be sent to attackers.	Hostile Individual Hostile Organization	High	High	Confid.-High Avail.-Mod.	SC-20, SC-21 *Note: Largely outside control of officials.*

Table 5: Threat Matrix for E-mail Registration and Ballot Request

quickly patched before any significant attack took place, and changes to the DNS system are being implemented to prevent similar attacks in the future [12].

However, there are less sophisticated attacks that could disrupt the election process. A denial of service attack could flood election officials with a massive number of fraudulent e-mails. The number of e-mails could quickly overwhelm the election official's e-mail server, preventing legitimate registration forms from reaching election officials. Denial of service attacks are very difficult to defend against, although filtering incoming e-mails could provide some protection. However, the resources necessary to carry out the attack are readily available to malicious individuals or groups, using roughly the same technology as systems that send large amounts of unsolicited e-mail (i.e. spam). Depending on the e-mail server settings, voters may or may not be automatically informed that their registration materials were discarded.

Current e-mail-based attacks on banking sites point to phishing as a likely attack on e-mail-based registration systems. That is, an attacker would contact a large number of voters, claiming to be their local election official and attempting to convince them to reply with their voter registration information. While a relatively small number of voters may be tricked into supplying their information, the attack could be conducted on a large scale. It is relatively easy and cheap to contact a very large numbers of voters, some of whom would almost certainly be fooled.

Digital signatures would provide an alternative method for authenticating voters. Voters with a public/private key pair could digitally sign their registration form, which could be verified by election officials upon receipt. Digital signatures would be nearly impossible to forge, and the process would not put sensitive personal information at risk of being intercepted. However, it would require a large-scale, potentially nation-wide, Public Key Infrastructure, which does not yet exist.

6.1.5 Web-Based

A web-based registration and ballot request system would perform voter authentication using secret personal information from the voter. However, unlike other systems, interception or modification in transit is not a significant threat. Any web-based system can and should incorporate encryption and integrity protection. All modern browsers ship with support for SSL/TLS [4,7], which is used extensively on e-commerce websites to provide such protections. Attackers may be able to intercept encrypted information in transit, but it is highly unlikely that they would be able to read or modify the protected information if web servers use properly configured implementations SSL/TLS.

While information in transit is secured, it would be possible to view voter information at the two end-points in the system: the voter's computer and the election web server. Malicious code, in the form of a computer virus or a Trojan horse, could record sensitive voter information and pass it to an attacker. Similarly, malicious individuals with access to the election web server could access sensitive voter information.

Attackers would be able to disrupt communications using denial of service attacks. A successful denial of service attack would overwhelm the election web server with traffic, preventing legitimate voters from sending registration and ballot request materials. It is very difficult to protect against denial of service attacks from an attacker with a large amount of resources. A successful denial of service attack generally requires access to a large number of computers with high-speed Internet connections. While an attacking organization may purchase these systems, it typically would use a Botnet. A Botnet is a collection of personal computers that have been infected with a virus that gives an attacker control of the computer. Control of Botnet-infected computers is sold on the black market, given nearly anyone with financial resources the technical resources to perform a denial of service attack.

Threat	Threat-Sources	Effort	Detection	Impact	Possible Controls
Ineligible individual allowed to register to vote.	Hostile Individuals	Low.	Mod.	Integrity-Mod.	IA-1, IA-2, IA-4, IA-5, IA-7
Voter information from registration/request materials is read or modified on thee election web server by authorized individuals.	Network Operators	Low	Low	Integrity-Mod.	AC-2,AC-3, AC-5, AC-6, SC-9, SC-12, SC-13
Voter information from registration/request materials is read or modified on thee election web server by unauthorized individuals.	Hostile Individuals Hostile Organizations	Mod.	Mod.	Confid.-High Integrity-High	AC-2,AC-3, AC-5, AC-6, AC-12, SC-9, SC-12, SC-13
A denial of service attack against the election web server overwhelms resources and prevents the transmission of registration/request materials.	Hostile Organizations	Mod.	High	Avail-High IR-4,	IR-5, CP-7, CP-8, SC-5
A denial of service attack against DNS servers disrupts access to the election web server	Hostile Organizations	High	High	Avail-High	*Outside control of officials.*
Election official offices have too few resources (e.g. bandwidth, servers) to handle legitimate traffic.	Network Operators Election Officials	Low	High	Avail-High IR-4,	IR-5
Sensitive personal information is intercepted between the voter and election official on the Internet.	Hostile Organizations Network Operators	High	Low	Confid.-High	PE-4, SC-6, SC-7, SC-12, SC-13
Malicious code (e.g. spyware) on the voter's computer transmits personal information from the registration/request materials to a third party.	Hostile Individual Hostile Organization	High	Mod.	Confid.-High	*Outside control of officials.*
Malicious code (e.g. a Trojan horse) on a voter's computer modifies or disrupts communication with the election web server.	Hostile Individual Hostile Organization	High	Mod.	Integrity-High Avail.-Mod.	*Outside control of officials.*
Defects in the election web server software causes voter information to be recorded incorrectly.	System Manufacturers	Mod.	Low	Integrity-High	SI-2, CM-2, CM-3, CM-5
Malicious code is inserted into the election web server which causes voter information to be recorded incorrectly.	Hostile Individual Hostile Organization	High	Mod.	Integrity-High	IA-2, AC-3, CM-5, MA-2, MA-3, MA-5, SI-3, SI-4, SI-7, PE-2, PE-3, PS-2, PS-3
Voters submit registration request materials to an incorrect web site (e.g. through phishing).	Hostile Individual Hostile Organization	Mod.	High	Confid.-Mod. Avail.-Mod.	*Largely outside control of officials.*
An attack on the DNS system forwards voters to an incorrect web site.	Hostile Individual Hostile Organization	High	High	Confid.-High Avail.-Mod.	SC-20, SC-21 *Note: Largely outside control of officials*

Table 6: Threat Matrix for Web-Based Registration and Ballot Request

However, the most likely threat for web-based registration processes comes from attackers that lure voters to fake websites posing at legitimate sites operated by election officials. This could be done via sophisticated technical attacks, or simple social engineering attacks. Internet web sites rely on DNS [11] to route traffic to the correct web server using a human-readable address. An attacker could trick one or more DNS servers into thinking that a fraudulent web server is a proper election web server. Voters attempting to navigate to their local election official's website could unknowingly navigate to a fake website, and supply attackers with sensitive personal information. Alternatively, an attacker could lure voters to a fake site by e-mailing them a link to a fraudulent web site. This is a common attack on Internet banking users.

Digital signatures would provide an alternative method for authenticating voters. Voters with a public/private key pair could digitally sign their registration form, which could be verified by election officials upon receipt. Digital signatures would be nearly impossible to forge, and the process would not put sensitive personal information at risk of being intercepted. However, it would require a large-scale, potentially nation-wide, Public Key Infrastructure, which does not yet exist.

6.2 Ballot Distribution

The section documents threats to the transmission options for the Ballot Distribution stage. This section discusses threats to systems which use postal mail, fax machines, electronic mail, and web sites to distribute blank ballots to registered UOCAVA voters. The systems analyzed in this section are discussed in Section 4.3. Note that telephone systems are not considered in this section. Telephone voting systems provide voters with ballot questions and allow voters to select their votes. Therefore, telephone voting systems are a type of ballot return system, and are discussed in Section 6.3.2.

6.2.1 Postal Mail

It is important for blank ballots to reach individual voters quickly and without modification. Postal mail is the slowest communications method considered in this paper. One of the greatest threats to postal mail delivery of ballots is not necessarily a malicious attack; it is that the unexpected delays in the postal mail system would cause ballots to be delivered too late to voters. Given transit times between many overseas locations and local election offices, it is unlikely that it would be possible to successfully recover from such delays.

Large scale malicious attacks are difficult to conduct on postal mail delivery of ballots. The only individuals capable of preventing the proper distribution of blank ballots to a large number of voters are election officials charged with operating the system. Smaller scale attacks on individual voters, or on a small number of voters are also possible, but their effect would be limited. Hostile individuals could steal blank ballots directly out of a voter's mailbox or place of residence, but this would not pose a major threat to the election as a whole.

Threat	Threat-Sources	Effort	Detection	Impact	Possible Controls
Individual delays or disrupts the process of preparing and/or mailing ballots.	System Operators Election Officials	Mod.	High	Avail.-High	PE-2, PE-3, PS-2, PS-3
Election official incorrectly indicates a voter is sent a ballot.	System Operators Election Officials	Mod.	Mod.	Avail.-Mod.	AC-2, AC-3, AC-5, AC-6, PS-2, PS-3
Election official sends a voter the wrong ballot.	Election Officials	Mod.	High	Avail.-Mod	AC-2, AC-3, AC-5, AC-6, PS-2, PS-3
Normal mail service fluctuations cause some ballots to be delivered late, or not at all.	Postal Workers	Low	High	Avail.-Mod.	MP-5, IR-4, IR-5
An attack disrupts mail service, causing some ballots to be delivered late, or not at all.	Postal Workers Hostile Individuals	Low	High	Avail.-Mod.	MP-5, IR-4, IR-5
Individual intercepts mailed ballots prior to being picked up by the intended recipient.	Hostile Individuals	Low	High	Avail.-Low	MP-5
Individual modifies electronic ballot file prior to ballot printing.	System Operators Election Officials	High	High	Integrity-Mod	AC-2, AC-3, AC-5, AC-6, PE-2, PE-3, PS-2, PS-3
Individual modifies paper ballots.	Election Officials	Mod.	High	Integrity-Mod	PE-2, PE-3, PS-2, PS-3, MP-1, MP-2, MP-4
Blank ballots are printed too late to reach voters on time.	System Operators Election Officials	Mod.	High	Avail.-High	IR-4, IR-5

Table 7: Threat Matrix for Postal Mail Ballot Delivery

6.2.2 Fax

Faxed distribution of blank ballots would not be subject to the same problems as the postal mail with delivery times. Faxed ballots would reach their destination nearly instantaneously. While it may be possible for an individual to eavesdrop on the faxed communications, this would only be a concern if blank ballots are accompanied by sensitive personal information about the voter.

Voters would not be able to predict the exact delivery time of their blank ballots. In many cases, ballots may be sent to a public fax machine, perhaps one shared by multiple employees at a

Threat	Threat-Sources	Effort	Detection	Impact	Possible Controls
Individual delays or disrupts the process of preparing and/or faxing ballots.	System Operators Election Officials	Mod.	High	Avail.-High	PE-2, PE-3, PS-2, PS-3
Election official incorrectly indicates a voter is sent a ballot.	System Operators Election Officials	Mod.	Mod.	Avail.-Mod.	AC-2, AC-3, AC-5, AC-6, PS-2, PS-3
Election official sends a voter the wrong ballot.	Election Officials	Mod.	High	Avail.-Mod.	AC-2, AC-3, AC-5, AC-6, PS-2, PS-3
An unauthorized individual takes a faxed ballot intended for a different voter.	Hostile Individuals	Low	Mod.	Avail.-Low	*Outside control of officials.*
A denial of service attack, or other technical attack, prevents outgoing faxes.	Telephone Operators Hostile Organizations	High	High	Avail.-Mod	IR-4, IR-5, SC-5, CP-7, CP-8, SC-13, SC-14
An individual modifies the paper ballots used by election officials prior to faxing a copy to a voter.	System Operators Election Officials	High	High	Integrity-Mod	PE-2, PE-3, PS-2, PS-3, MP-1, MP-2, MP-4

Table 8: Threat Matrix for Fax Ballot Delivery

workplace. Blank ballots may remain in the fax machine for an extended period of time before being noticed by the intended recipient. This would provide would-be attackers with ample opportunities to intercept the ballot before it reaches the intended recipient. While it would be very difficult for a single individual to intercept a large number of blank ballots, there are some situations where this might be possible. A single individual at a military base may collect and distribute faxes for a large number of soldiers stationed at the base.

Faxed ballots have little integrity protection in transit. However, it is quite difficult to modify faxes in transit, so this is not a significant threat. A more serious threat is that ballots could be modified prior to being faxed by malicious election employees, or after being sent to the recipient's fax machine. Voters may be able to detect changes to the ballot if certain ballot questions have been left off or modified.

6.2.3 Electronic Mail

E-mailed ballots would not be subject to the same problems as the postal mail with delivery times. Like faxed ballots, e-mailed ballots would reach their destination nearly instantaneously. Eavesdropping is a potential threat whenever Internet communications are involved, and particularly with e-mailed communications, which are sent unencrypted. However, as ballot contest information need not be secret, eavesdropping is only a significant threat if ballots are accompanied by sensitive personal information about the voter.

E-mails are significantly easier to intercept and modify in transit than other forms of communication. E-mails travel through telecommunications lines, network equipment and e-mail servers before reaching the intended recipient. Anyone with access to the infrastructure could read or even modify e-mail messages. In particular, e-mail servers often store messages for a short period of time before passing them on to the next server, or the intended recipient. System operators for these servers would be in a good position to intercept or modify e-mailed ballots. Voters may be able to detect any changes made to the blank ballot. In addition, certain technical measures could be taken to assist voters in identifying improperly modified ballots.

Denial of service attacks are possible against election official e-mail servers, but very difficult to conduct. While it is comparatively easy to prevent an individual or organization from receiving an e-mail, it is much more difficult to stop a message from being sent. While blank ballot delivery is time-sensitive, the acceptable time frame window is several days. This would likely provide election officials with a sufficient amount of time to recover from any denial of service attack and distribute blank ballots on time.

Threat	Threat-Sources	Effort	Detection	Impact	Possible Controls
Individual delays or disrupts the process of preparing and/or e-mailing ballots.	System Operators Election Officials	Mod.	High	Avail.-High	PE-2, PE-3, PS-2, PS-3
Election official incorrectly indicates a voter is sent a ballot.	System Operator Election Official	Mod.	Mod.	Integrity-Mod.	AC-2, AC-3, AC-5, AC-6, PS-2, PS-3
Election official sends a voter the wrong ballot.	Election Official	Mod.	High	Integrity-Mod.	AC-2, AC-3, AC-5, AC-6, PS-2, PS-3
An unauthorized individual gains access to the voter's computer and/or e-mail accounts and accesses the blank ballot.	Hostile Individual	Mod.	Low	Confid.-Low	*Outside control of officials.*
Ballot files are modified on the e-mail servers of the voter or election official by authorized system administrators.	System Operators Election Officials	Mod.	Low	Integrity-Mod.	PS-2, PS-3, AC-3, AC-5, AC-6, SC-8
Ballot files are modified on the e-mail servers of the voter or election official by unauthorized individuals.	Hostile Individual Hostile Organization	High	Low	Integrity-Mod.	AC-3, AC-5, AC-6, IR-4, IR-5, SC-7, SC-8, SI-5
A denial of service attack against voter and/or election official e-mail servers overwhelms resources and prevents the transmission of blank ballots.	Hostile Organization Network Operators	High	High	Avail.-High	IR-4, IR-5, SC-5, CP7, CP-8, SC-14
Election official offices have too few resources (e.g. bandwidth, servers) to handle legitimate traffic.	Network Operators Election Officials	Low	High	Avail-High	IR-4, IR-5
A voter receives a spoofed e-mail with an improper blank ballot or instructions, and assumes it is proper.	Hostile Individual Hostile Organization	Low	High	Integrity-High	SC-8, SC-13, SC-14
Malicious code (e.g. a Trojan horse) on a voter's computer modifies the received ballot or prevents the proper delivery of the ballot.	Hostile Individual Hostile Organization	High	Mod.	Integrity-High Avail.-High	*Outside control of officials.*
An attack on the DNS system prevents ballots from reaching their intended recipients.	Hostile Individual Hostile Organization	Mod.	High	Avail.-Mod.	IR-4, IR-5, SC-20, SC-21

Table 9: Threat Matrix for E-mail Ballot Delivery

6.2.4 Web-Based

Web-based communication can be easily protected using properly configured SSL/TLS, virtually eliminating the threat of eavesdropping or ballot modification in transit. Some attacks could take place at the endpoints: on the election web server and on voters' computers. A malicious election official could load improper ballots on the web site, although this would likely be quickly detected and resolved. Smaller scale attacks could take place on voters' computers. A hostile individual with access to a voter's computer could modify already downloaded ballots.

A significant threat to web-based ballot distribution is that attackers could lure voters to fake web sites posing as legitimate sites operated by election officials. This could be done via sophisticated technical attacks, or simple social engineering attacks. Internet web sites rely on DNS to route traffic to the correct web server using a human-readable address. An attacker could trick one or more DNS servers into thinking that a fraudulent web server is a proper election web server. Voters attempting to navigate to their local election official's website could unknowingly find themselves on a fake website. Voters may provide their voter credentials on this web site, potentially allowing the attacker to impersonate them in future transactions. Voters

Threat	Threat-Sources	Effort	Detection	Impact	Possible Controls
Individual delays or disrupts the process of preparing ballots or uploading them to the election web server.	System operators Election officials	Mod.	High	Avail.-Mod.	PE-2, PE-3, PS-2, PS-3
An unauthorized individual gains access to the voter's computer and accesses an already-downloaded blank ballot.	Hostile Individuals	Mod.	Low	Confid.-Low	Outside control of officials.
An unauthorized individual downloads a blank ballot intended for a different voter by gaining improper access to the election web server.	Hostile Individuals	Mod.	Low	Integrity-Mod. Avail.-Mod.	AC-2, AC-3, IA-2, SC-7, SI-4
Blank ballots are modified on the election web servers by authorized system administrators.	System operators Election officials	Mod.	Low	Integrity-Mod.	PE-2, PE-3, PE-6, PS-2, PS-3, AU-2, AU-3, AU-4, AU-6, AU-7, AU-8, AU-9, AU-10, AC-2, AC-3, AC-5, AC-6, SC-8, SC-13
Blank ballots are modified on the election web servers by unauthorized individuals with physical access to the server.	Hostile Individuals	High	Mod.	Integrity-Mod.	AC-2, AC-3, IA-2, PE-2, PE-3, PE-5, PE-6, PS-2, PS-3, SC-8, SC-13
Attackers remotely access election web servers and modify blank ballots.	Hostile Individuals	High	Mod.	Integrity-Mod.	AC-2, AC-3, IA-2, SC-7, SI-4
A denial of service attack against voter and/or election official e-mail servers overwhelms resources and prevents the transmission of blank ballots.	Hostile Organizations	Mod.	High	Avail-High	IR-4, IR-5, CP-7, CP-8, SC-5
Election official offices have too few resources (e.g. bandwidth, servers) to handle legitimate traffic.	Network Operators Election Officials	Low	High	Avail-Mod.	IR-4, IR-5
A voter is tricked into going to a spoofed site to download a fake ballot.	Hostile Individual Hostile Organization	Low	High	Integrity-High	Largely outside control of officials.
An attack on the DNS system forwards voters to an incorrect website.	Hostile Organizations	High	High	Avail-High	SC-20, SC-21 Note: Largely outside control of officials.
Malicious code (e.g. a Trojan horse) on a voter's computer modifies the received ballot or prevents the proper delivery of the ballot.	Hostile Individual Hostile Organization	High	Mod.	Integrity-High Avail.-High	Outside control of officials.

Table 10: Threat Matrix for Web-Based Ballot Delivery

may also download improper ballots that, if marked and returned, would have to be thrown out by election officials. Alternatively, an attacker could lure voters to a fake site by e-mailing them a link to a fraudulent web site. This is a common attack on Internet banking users.

Denial of service attacks are a significant threat to any web-based ballot distribution mechanism. A successful denial of service attack would overwhelm the election web server with traffic, preventing legitimate voters from obtaining blank ballots. As previously noted, it is very difficult to protect against denial of service attacks from an attacker with a large amount of resources. A successful denial of service attack generally requires access to a large number of computers with high-speed Internet connections, but such resources could be easily obtaining by buying time on a Botnet.

Malicious code on voters' computers could prevent them from successfully downloading a ballot. A computer virus could prevent a voter from reaching the election web site, or it could even redirect the voter to an attacker's fraudulent web site. Voters who do not detect the fraudulent site might enter their voter credentials on the site, potentially allowing the attacker to impersonate those voters in future transactions.

6.3 Ballot Return

The section documents threats to the transmission options for the return of ballots. This section discusses threats to systems which use postal mail, telephones, fax machines, electronic mail, and web sites to allow voters to submit votes to their jurisdictions. The systems analyzed in this section are discussed in Section 4.4.

6.3.1 Postal Mail

Returning voted ballots is a very time-sensitive task. Many voters do not receive blank ballots until very close to the Election Day, which does not give them a lot of time to vote and return the ballot. Most states have deadlines for when absentee ballots must be postmarked and delivered to election offices. Malicious postal workers may be able to selectively identify absentee ballots in the mail, and disrupt delivery. However, typically a single employee would not encounter enough absentee ballots to pose a significant threat to the election outcome, except, for example, on a military base where a single solider handles all outgoing mail. Hostile organizations may be able to attack sorting facilities or transports. Such attacks would be very dangerous and difficult to conduct, and the likely number of ballots affected is small. However, normal fluctuations in delivery times could affect a large number of voters, delaying their ballots long enough to cause them to miss deadlines imposed by states.

Threat	Threat-Sources	Effort	Detection	Impact	Possible Controls
Impersonation of registered voter (e.g., forged signature).	Hostile Individuals	Mod.	Mod.	Integrity-Mod.	IA-1, IA-2, IA-4, IA-5,
Voter coerced into voting a particular way.	Hostile Individuals	Low	Mod.	Confid.-Mod.	*Outside control of officials.*
Normal mail service fluctuations cause some ballots to be returned late, or not at all.	Postal workers	Low	High	Avail.-Mod.	MP-5
An attack disrupts mail service, causing some ballots to be returned late, or not at all.	Postal workers, Hostile Organizations	High	High	Avail.-Mod.	MP-5
A large-scale attack on a postal mail hub disrupts mail delivery for a large group of voters.	Hostile Organizations	High	High	Avail.-High.	MP-5, IR-4, IR-5
Sensitive voter information is intercepted from the ballot while it is in the mail.	Postal workers, System operators, Election officials	Mod.	Low	Confid.-Low	MP-5
Marked ballots are modified or destroyed at the election office.	System operators, Election officials	High	Mod.	Integrity-Mod.	MP-1, MP-2 (1), MP-4, PE-2, PE-3 (1), PS-2, PS-3
Marked ballots are viewed by unauthorized personnel, resulting in loss of voter privacy.	System operators, Election officials, Postal worker	High	Low	Confid.-Mod.	MP-1, MP-2 (1), MP-4, PE-2, PE-3 (1), PS-2, PS-3
Election officials are flooded with a large number of illegitimate ballots.	Hostile Organizations	Mod.	High	Avail-Mod. Integrity-Mod.	MP-2

Table 11: Threat Matrix for Postal Mail Ballot Return

Confidentiality is important during the ballot return stage of the voting process. At a minimum, a ballot will show a voter's selections on the ballot questions. In some cases, the ballot may be accompanied by sensitive personal information about the voter. While postal employees and hostile organizations may be able to intercept and read a small number of ballots, the overall effect on the election would be quite small. It is difficult to imagine a large scale loss of personal information during transmission through the postal service.

Voted ballots are at higher risk before and after transmission through the mail. Hostile individuals could steal a ballot from a legitimate voter, forge the voter's signature and return the voted ballot to the election official. Alternatively, a hostile individual could coerce a voter into voting for a particular candidate. In either case, a single hostile individual or organization would be limited in the number of votes they could steal or unduly influence. There is far more potential to influence or damage an election at the election official's offices. There, a large number of voted ballots would be collected and stored for several days or weeks. Hostile individuals with physical access to these ballots could violate voter secrecy, modify ballots, or destroy ballots. Tight physical access controls could reduce, but not eliminate, this threat.

6.3.2 Telephone

Telephone voting would virtually eliminate delays caused by ballot distribution and return. The voter would be given a set of ballot options and immediately be allowed to select his or her

Threat	Threat-Sources	Effort	Detection	Impact	Possible Controls
Impersonation of registered voter (e.g. stolen PIN).	Hostile Individuals	Mod.	Mod.	Integrity-Mod.	IA-1, IA-2, IA-4, IA-5, IA-7
Voter coerced into voting a particular way.	Hostile Individuals Hostile Organizations	Low	Mod	Confid.-Mod	*Outside control of officials.*
Election official offices have too few telephone lines to handle demand.	Telephone Operators System Operators	Low	High	Avail.-High	IR-4, IR-5
A denial of service attack against the election official office jams telephone lines.	Telephone operators Hostile Organizations	Mod.	High	Avail.-High	IR-4, IR-5, CP-7, CP-8, SC-5
Sensitive personal information or ballot selections are intercepted en route.	Telephone Operators Hostile Organizations	High	Low	Confid.-Mod.	PE-4, SC-8, SC-9, SC-12, SC-13
Voter ballot selections are viewed on the server by individuals with authorized access to the election system, resulting in loss of voter privacy.	Election Official System Operators	Mod.	Low	Confid. High.	PE-2, PE-3, PE-6, PS-2, PS-3, AU-2, AU-3, AU-4, AU-6, AU-7, AU-8, AU-9, AU-10, AC-2, AC-3, AC-5, AC-6
Voter ballot selections are viewed on the server by unauthorized personnel, resulting in loss of voter privacy.	Hostile Individuals	High	Mod.	Confid.-High	AC-2, AC-3, IA-2, PE-2, PE-3, PE-5, PE-6, PS-2, PS-3
Defects in the voting system server software cause votes to be recorded incorrectly.	System Manufacturers	Mod.	Low	Integrity-High	SI-2, CM-2, CM-3, CM-5
Malicious code is inserted into the voting system server software which causes votes to be recorded incorrectly.	Election Official System Operators	Mod.	Low	Integrity-High	IA-2, AC-3, CM-5, MA-2, MA-3, MA-5, SI-3, SI-4, SI-7, PE-2, PE-3, PS-2, PS-3
An attacker tricks voters into calling the wrong phone number to vote.	Hostile Individual Hostile Organization	Low	High	Integrity-High	*Largely outside control of officials.*

Table 12: Threat Matrix for Telephone Ballot Return

choices. As noted in Section 6.1.2, it may be possible for hostile individuals with access to the telephone network infrastructure to eavesdrop on or disrupt these telephone calls. The threat is increased in the case of cellular phone communications. In general, however, a successful large-scale attack would be needed to target the communications equipment close to the election office housing the telecommunications equipment. This would substantially reduce the number of individuals capable of conducting an attack.

Sabotaging the telephone network equipment, or jamming the telephone lines, would require a comparable amount of access to network equipment, but would be significantly easier to conduct, particularly in the case of jamming cellular phone communications. Such an attack would prevent legitimate voters from accessing the equipment necessary to cast a ballot. Attackers could also conduct a denial of service attack on the telephone voting system by continuously calling and tying up communications lines. This would also prevent legitimate voters from casting a ballot.

Most telephone systems could feature an automated calling center capable of interacting with the voter similar to those used by many businesses. Election officials would not need to physically handle voted ballots, but would have access to the information stored on the server. While access control mechanisms could restrict access to this information, any hostile individual capable of bypassing these controls could change or delete a large number of ballots. A sophisticated attacker may be able to make these changes without leaving any evidence in, for example, the system event log.

Automated telephone voting is a form of electronic voting. The computer system running the automated calling center would have to be trusted to accurately record voters' selections. Defects in the voting system software, or malicious code installed on the voting system by hostile individuals, could cause votes to be recorded improperly, or could modify votes at a later time.

As noted in Section 6.1.2, some individuals and organizations are using Voice-over-Internet-Protocol (VoIP) telephones, which transmit information over the Internet instead of the public telephone network. Use of the Internet to transmit their ballot selections and choices would substantially increase the risk of eavesdropping and modification attacks in-transit. Such systems would be subject to many of the risks associated with e-mail and web-based Internet voting.

6.3.3 Fax

Faxed ballot return is an alternative to mailing ballots. A fax-based system for returning ballots would not experience problems with delays. However, certain election officials would have the necessary level of access to compromise voter secrecy, and potentially to modify votes. Faxed ballots could remain in the fax machine for some period of time before being placed in a secure ballot box. Individuals with access to the fax machine or the ballot box would be in a position to violate voter privacy by accessing these ballots. Also they might be able to replace the faxed ballots with other ballots containing different votes.

Threat	Threat-Sources	Effort	Detection	Impact	Possible Controls
Impersonation of registered voter (e.g. forged signature).	Hostile Individuals	Mod.	Mod.	Integrity-Mod.	IA-1, IA-2, IA-4, IA-5, IA-7
Voter coerced into voting a particular way.	Hostile Individuals Hostile Organizations	Low	Mod	Confid.-Mod	*Outside control of officials.*
Election official offices have too few fax machines and/or telephone lines to handle demand.	Telephone Operators System Operators	Low	High	Avail.-High	IR-4, IR-5
A denial of service attack against the election official office jams fax machines and/or telephone lines.	Telephone Operators Hostile Organizations	Mod.	High	Avail.-High	IR-4, IR-5, CP-7, CP-8, SC-5
Personally identifiable material is intercepted en route.	Telephone Operators Hostile Organizations	High	Low	Confid.-Mod.	PE-4, SC-8, SC-9, SC-12, SC-13
Election officials are flooded with a large number of illegitimate faxed ballots.	Hostile Organizations	Mod.	High	Avail-Mod. Integrity-Mod.	IR-4, IR-5
An attacker tricks voters into calling the wrong phone number to vote.	Hostile Individual Hostile Organization	Low	High	Integrity-High	*Outside control of officials.*
Disgruntled election official fails to properly handle faxed ballots.	Election Official	Mod.	Low	Integrity-Mod.	PS-2, PS-3
Sensitive personal information and/or ballot selections are improperly read from faxed votes.	Election Officials Support Staff Hostile Individuals	Mod.	Mod.	Confid.-Mod	PE-2, PE-3, PE-5, PE-6, PS-2, PS-3
Sensitive personal information and/or ballot selections are improperly read from received ballots in storage.	Election Officials	Mod.	Mod.	Confid.-High	MP-1, MP-2, MP-4, PE-2, PE-3, PE-6, PS-2, PS-3
Electronic copies of faxed ballots are read of the memory of fax machines.	Hostile Individuals	High	Low	Confid.-Mod.	*Largely outside the control of officials.*

Table 13: Threat Matrix for Fax Ballot Return

Denial of service attacks may be possible against these systems. Malicious groups could flood election fax machines with large numbers of illegitimate ballots. Such an attack would have two major results. First, the illegitimate traffic could tie up communication lines, preventing legitimate voters from casting ballots. Second, it may be difficult for election officials to distinguish the legitimate ballots from the illegitimate ballots. Postal mail distribution and return of ballots could limit the number of forged ballots since valid inbound ballots would need to be on the proper paper stock; however, there would be no such protection with faxed ballots. Illegitimate votes would have to be identified using the voter identification and authentication information (e.g. a voter's signature), possibly with the assistance of any ballot tracking information. A small number of illegitimate ballots may be able to pass through these checks.

6.3.4 Electronic Mail

In most instances, voted ballots returned via e-mail would reach election officials nearly instantaneously. Communications could, however, be disrupted by malicious parties. Denial of service attacks are a significant threat to e-mail-based voting systems. Attackers could flood election e-mail servers with large amounts of illegitimate traffic. This could not only prevent voters' e-mails from reaching election officials, but could also make it difficult for officials to distinguish between valid and invalid ballots.

Eavesdropping is a potential threat whenever Internet communications is involved, and particularly with e-mailed communications, which are sent unencrypted. While eavesdropping is not a significant threat for ballot distribution, as that information is generally publically available, voted ballots must remain confidential. Voted ballots show how an individual voted, and may sometimes contain sensitive personal information about the voter. E-mails are significantly easier to intercept and modify in transit than other forms of communication. E-mails travel through telecommunications lines, network equipment and e-mail servers before reaching the intended recipient. Anyone with access to the infrastructure could read or even modify e-mail messages. In particular, e-mail servers often store messages for a short period of time before passing them on to the next server, or the intended recipient. System operators for these servers could intercept or modify e-mailed ballots. It is unlikely that election officials would be able to identify ballots that had been modified in-transit.

Also, e-mailed ballots are at risk before and after they are sent to election officials. Voters' computers could be infected with malicious code capable of disrupting communications with an election official. Very sophisticated attacks may be able to modify digital ballots prior to e-mailing them to election officials. Malicious code would need to spread to a large number of personal computers before it would have a substantial effect on an election. The computer virus may be detected before election day, but there would be no way for election officials to identify affected ballots. Similar malicious code on election computer systems could have the same effect.

E-mail does not provide any guarantee that the intended recipient will receive the message. The e-mail system relies on the DNS system [11] to route e-mails to the proper servers. An attack on DNS servers could route e-mails to an attacking party. This would not only result in voter disenfranchisement, but also the loss of sensitive voter information. This kind of attack would require very sophisticated attackers focusing their efforts on major e-mail service providers. There are no known reports of a similar attack being successfully conducted on e-mail or DNS servers. However, it is important to note that a recent vulnerability was discovered in DNS servers that could have been used to construct a similar attack [13]. DNS servers were quickly patched before any significant attack took place.

Less sophisticated, but equally effective, attacks may attempt to trick voters into sending their ballots to an attacker. That is, an attacker would contact a large number of voters, claiming to be their local election official and attempting to convince them to reply with their cast ballot. While a relatively small number of voters may be fooled, it is relatively easy and cheap to contact a very large numbers of voters.

Threat	Threat-Sources	Effort	Detection	Impact	Possible Controls
Impersonation of registered voter (e.g., forged signature)	Hostile Individuals	Mod.	Mod.	Integrity-Mod.	IA-1, IA-2, IA-4, IA-5, IA-7
Voter coerced into voting a particular way.	Hostile Individuals Hostile Organizations	Low	Mod	Confid.-Mod	*Outside control of officials.*
A denial of service attack against voter and/or election official e-mail servers overwhelms resources and prevents the transmission of voted ballots.	Hostile Organizations	Mod.	High	Avail-High	IR-4, IR-5, CP-7, CP-8, SC-5
Election official offices have too few resources (e.g., bandwidth, servers) to handle legitimate traffic.	Network Operators Election Officials	Low	High	Avail-High	IR-4, IR-5
Sensitive personal information or ballot selections are intercepted between the voter and election official on the Internet.	Hostile Organizations Network Operators	High	Low	Confid.-Mod	PE-4, SC-9, SC-12, SC-13
Voted ballots are modified while being transmitted to the election official (e.g. on e-mail servers)	Hostile Organizations Network Operators	High	Low	Integrity.-Mod	SC-8, SC-12, SC-13
Malicious code (e.g., a Trojan horse) on a voter's computer modifies or disrupts outgoing e-mails containing voted ballots.	Hostile Individuals Hostile Organizations	High	Mod.	Integrity.-Mod	*Outside control of officials.*
Voter ballot selections are accessed off election information systems by individuals with authorized access to these machines, resulting in loss of voter privacy.	System Operators Election Officials	Mod.	Low	Confid.-High	PE-2, PE-3, PE-6, PS-2, PS-3, AU-2, AU-3, AU-4, AU-6, AU-7, AU-8, AU-9, AU-10, AC-2, AC-3, AC-5, AC-6
Voter ballot selections are accessed off election information systems by unauthorized personnel, resulting in loss of voter privacy.	Hostile Individuals Election Officials	High	Low	Confid.-High	AC-2, AC-3, IA-2, PE-2, PE-3, PE-5, PE-6, PS-2, PS-3
Individuals with physical access to election information systems delete or modify ballots stored on these systems	Hostile Individual System Operators Election Official	High	Low	Integrity-High	AC-2, AC-3, IA-2, PE-2, PE-3, PE-5, PE-6, PS-2, PS-3
Unauthorized individuals remotely access election information systems and view, modify or delete ballots stored on these systems	Hostile Individuals Hostile Organizations	High	Mod	Confid.-High Integrity-High	AC-2, AC-3, IA-2, SC-7, SC-8, SC-13, SI-4
Malicious code (e.g. a Trojan horse) on e the voter's-mail server modifies or deletes e-mails containing voted ballots.	Hostile Individuals Hostile Organizations	High	Low	Integrity-Mod.	*Outside control of officials.*
Malicious code (e.g. spyware) on the voter's e-mail server transmits voter ballot selections to a third party.	Hostile Individuals Hostile Organizations	High	Low	Confid.-Mod.	SC-9, SC-13 *Largely outside control of officials.*
Malicious code (e.g., a Trojan horse) on the election official's e-mail server modifies or deletes e-mails containing voted ballots.	Hostile Individuals Hostile Organizations	High	Mod.	Integrity-High	IA-2, AC-3, CM-3, CM-5, MA-2, MA-3, MA-5, SI-3, SI-4, SI-7, PE-2, PE-3, PS-2, PS-3, SC-7, SC-8, SC-13
Malicious code (e.g., spyware) on the election official's e-mail server transmits voter ballot selections to a third party.	Hostile Individuals Hostile Organizations	High	Mod.	Confid.-High	IA-2, AC-3, CM-3, CM-5, MA-2, MA-3, MA-5, SI-3, SI-4, SI-7, PE-2, PE-3, PS-2, PS-3, SC-7, SC-9, SC-13
Disgruntled election officials fail to properly record the e-mailed vote.	Election Official	Mod.	Low	Integrity-Mod.	PS-2, PS-3
An individual reads, modifies or destroys an e-mailed ballot in storage, after it has been printed, but before being tallied.	Election Official	Mod.	Mod.	Confid.-High Integrity-High	PS-2, PS-3, PE-2, PE-3, MP-1, MP-2, MP-4
Voters are tricked into sending voted ballots to an incorrect e-mail address, resulting in the disenfranchisement and the loss of personal information.	Hostile Individual Hostile Organization	Low	High	Confid.-High Avail.-Mod.	*Outside control of officials.*
An attack on the DNS system causes e-mails containing voted ballots to be sent to attackers.	Hostile Individual Hostile Organization	Mod.	High	Confid.-High Avail.-Mod.	*Largely outside control of officials.*

Table 14: Threat Matrix for E-mail Ballot Return

6.3.5 Web-Based

Web-based Internet voting is a form of electronic voting. The election web server would need to be trusted to accurately record voters' selections. Defects in the voting system software, or malicious code installed on the voting system by hostile individuals, could cause votes to be recorded improperly, or could modify votes at a later time. Skilled hackers may find vulnerability in the voting system software that would grant them access to voter and ballot information. This could also lead to a loss of voter secrecy, or a loss of election integrity. Sophisticated attacks would leave little or no evidence.

Election officials, or other individuals with physical access to voting system equipment, may be able to gain access to election information, including cast ballots. Sophisticated attackers may also be able to delete any audit records that would leave evidence of their attack.

Denial of service attacks are significant threats to Internet-based voting systems. A successful denial of service attack would overwhelm the election web server with traffic, preventing legitimate voters from casting a ballot. It is very difficult to protect against denial of service attacks from an attacker with a large amount of resources. A successful denial of service attack generally requires access to a large number of computers with high-speed Internet connections. While an attacking organization may purchase these systems, it typically would use a Botnet. A Botnet is a collection of personal computers that have been infected with a virus that gives an attacker control of the computer. Control of Botnet-infected computers is sold on the black market, given nearly anyone with financial resources the technical resources to perform a denial of service attack.

Many of the potential threats to a web-based Internet voting system involve attacks on equipment that are not under election officials' control. Attacks on the DNS system could lead voters to fraudulent web sites. These voters may unknowingly provide their voter credentials to a malicious party, who in turn could impersonate the voter on the legitimate election server. Malicious code installed on voters' personal computers could disrupt communications with an election web server, or even modify voters' ballot choices without their knowledge. A computer virus would have to spread to a large number of computers before it could have a substantial effect on an election. Antivirus vendors may be able to identify and offer protections against such viruses, but not until after some voters' computers have been compromised. Furthermore, election officials would have no guarantee that their constituents would use updated anti-virus software. Election officials would have little recourse but to assume that all received votes are valid, as there would be no way to identify ballots from compromised machines.

Less sophisticated attackers may be able to trick voters into navigating to a fraudulent web site that would mimic the actual election site. This type of attack, known as phishing, involves sending a large number of messages to potential voters claiming to be from election officials. The message could instruct voters to log into the fraudulent web site to cast a ballot. While most voters would discard such messages, a small percentage of voters could fall victim to this attack, which is common in the banking industry.

Threat	Threat-Sources	Effort	Detection	Impact	Possible Controls
Impersonation of registered voter.	Hostile Individuals	Mod.	Mod.	Integrity-Mod.	IA-1, IA-2, IA-4, IA-5, IA-7
Voter coerced into voting a particular way.	Hostile Individuals, Hostile Organizations	Low	Mod	Confid.-Mod	*Outside control of officials.*
A denial of service attack against the election web servers overwhelms resources and prevents the transmission of voted ballots.	Hostile Organizations	Low	High	Avail-High	IR-4, IR-5, CP-7, CP-8, SC-5
Election official offices have too few resources (e.g. bandwidth, servers) to handle legitimate traffic.	Network Operators, Election Officials	Low	High	Avail-High	IR-4, IR-5
Personal information is intercepted between the voter and election official on the Internet.	Hostile Organizations, Network Operators	High	Low	Confid.-High	PE-4, SC-6, SC-7, SC-12, SC-13
Malicious code (e.g., a Trojan horse) on a voter's computer modifies communication with the election web server, modifying voted ballots before passing them to the server.	Hostile Individual, Hostile Organization	High	Mod.	Integrity-Mod.	*Outside control of officials.*
Malicious code (e.g., a Trojan horse) on a voter's computer disrupts communication with the election web server, preventing ballot return.	Hostile Individual, Hostile Organization	High	Mod.	Avail.-Mod.	*Outside control of officials.*
Voter ballot selections are accessed off election information systems by individuals with authorized access to these machines, resulting in loss of voter privacy.	System Operators, Election Officials	Mod.	Low	Confid.-High	PE-2, PE-3, PE-6, PS-2, PS-3, AU-2, AU-3, AU-4, AU-6, AU-7, AU-8, AU-9, AU-10, AC-2, AC-3, AC-5, AC-6
Voter ballot selections are accessed off election information systems by unauthorized personnel, resulting in loss of voter privacy.	Hostile Individuals, Election Officials	High	Low	Confid.-High	AC-2, AC-3, IA-2, PE-2, PE-3, PE-5, PE-6, PS-2, PS-3
Individuals with physical access to election information systems delete or modify ballots stored on these systems.	Hostile Individual, System Operators, Election Official	High	Low	Integrity-High	AC-2, AC-3, IA-2, PE-2, PE-3, PE-5, PE-6, PS-2, PS-3
Unauthorized individuals remotely access election information systems and view, modify or delete ballots stored on these systems.	Hostile Individuals, Hostile Organizations	High	Mod	Confid.-High, Integrity-High	AC-2, AC-3, IA-2, SC-7, SI-4
Malicious code (e.g. a Trojan horse) on the election web server deletes or modifies voted ballots.	Hostile Individual, Hostile Organization	High	Mod.	Integrity-High	IA-2, AC-3, CM-5, MA-2, MA-3, MA-5, SI-3, SI-4, SI-7, PE-2, PE-3, PS-2, PS-3
Malicious code (e.g., spyware) on the election web server transmits voter ballot selections to a third party.	Hostile Individual, Hostile Organization	High	Mod.	Confid.-High	IA-2, AC-3, CM-5, MA-2, MA-3, MA-5, SI-3, SI-4, SI-7, PE-2, PE-3, PS-2, PS-3
Malicious code (e.g., spyware) on a voter's computer transmits voter ballot selections to a third party.	Hostile Individual, Hostile Organization	High	Mod.	Confid.-Mod.	*Outside control of officials.*
Defects in the voting system server software cause votes to be recorded incorrectly.	System Manufacturers	Mod.	Low	Integrity-High	SI-2, CM-2, CM-3, CM-5
Voters are tricked into returning voted ballots via an incorrect web site (e.g. through Phishing), resulting in the disenfranchisement and the loss of personal information.	Hostile Organizations	Low	High	Integrity-High, Confid.-High	*Outside control of officials.*
An attack on the DNS system forwards voters to an incorrect website, resulting in the disenfranchisement and the loss of personal information.	Hostile Organizations	High	High	Integrity-High	*Largely outside control of officials.*

Table 15: Threat Matrix for Web-Based Ballot Return

7 Security Controls

The threat analysis conducted and documented in Section 6 includes references to security controls. These controls provide procedural and technical countermeasures to protect the confidentiality, integrity and availability of systems from threats. Whenever possible, specific controls are referenced for each threat identified in the analysis. These controls fully or partially mitigate the associated threat. In some cases the controls are preventative. That is, the controls prevent a security violation from taking place. In other cases the controls are reactive, in that they help recover from an attack or other security violation without further loss of confidentiality, integrity or availability. Preventative controls are preferable, but not always possible or realistic.

This section summarizes the security controls identified to mitigate threats to each transmission option. These controls point to specific controls listed in NIST SP 800-53, *Recommended Security Controls for Federal Information Systems* [3]. NIST SP 800-53 is a catalog of high-level security controls, written primarily for federal computer systems. This report references the controls documented in NIST SP 800-53 by the Control Number. As the controls are high-level, and not geared for election systems, this report includes discussion on how these controls could be implemented in UOCAVA election systems.

The particular security controls referenced in this report mitigate specific threats identified to each transmission option. Furthermore, threats are identified for the high-level characterizations of election systems outlined in Section 4. Most jurisdictions will use a variation of one or more of the systems identified in this paper. As such, specific voting systems may be vulnerable to different threats, requiring a different set of security controls. This report does not suggest that the following controls adequately mitigate the threats faced by each system. Individual jurisdictions should use threats and security controls in this report, along with specific information about their own systems and accompanying procedures, to ensure adequate security controls are in place. Furthermore, election systems should be designed with good security engineering principles, which may dictate additional security controls than those specified here. For instance, auditing functionality, an important component of any secure computer system, may not effectively mitigate any specific threat on its own, but it would provide useful information when responding to malicious attacks or simple malfunctions.

7.1 Postal Mail

Ctrl. Name	Stages RBR[1]	BD[2]	BR[3]	Control Text	Notes
AC-2	X			ACCOUNT MANAGEMENT	Databases and IT systems used to manage registration information should be protected with access control mechanisms.
AC-3	X			ACCESS ENFORCEMENT	
AC-5	X			SEPARATION OF DUTIES	
AC-6	X			LEAST PRIVILEGE	
IA-1	X	X		IDENTIFICATION AND AUTHENTICATION POLICY AND PROCEDURES	Officials should develop procedures and implement technical mechanisms to identify voters, election officials and system administrators. Procedures may be used to authenticate voters, while IT systems should include IA and access control mechanisms.
IA-2	X	X		USER IDENTIFICATION AND AUTHENTICATION	
IA-4	X	X		IDENTIFIER MANAGEMENT	
IA-5	X	X		AUTHENTICATOR MANAGEMENT	
IR-4	X	X		INCIDENT HANDLING	Officials should monitor for disruptions in their IT systems and in external essential systems, such as postal mail delivery.
IR-5	X	X		INCIDENT MONITORING	
MP-1	X	X	X	MEDIA PROTECTION POLICY AND PROCEDURES	Officials should protect registration forms, blank paper ballots and voted ballots with procedures. Care should taken when transporting these materials, both internally and via the postal service.
MP-2	X	X	X	MEDIA ACCESS	
MP-4	X	X	X	MEDIA STORAGE	
MP-5	X	X	X	MEDIA TRANSPORT	
PE-2	X	X	X	ACCESS CONTROL FOR TRANSMISSION MEDIUM	Officials should control physical access to vital systems and sensitive information.
PE-3	X	X	X	MONITORING PHYSICAL ACCESS	
PS-2	X	X	X	POSITION CATEGORIZATION	Screen election employees who will be handling registration forms and ballots.
PS-3	X	X	X	PERSONNEL SCREENING	

Access Control (AC):
IT systems containing important election information, such as an electronic voter registration database should be protected by access control mechanisms. Access to these systems should be limited to employees who need this information to perform election-related duties. Furthermore, individuals who need access to some voter information should not necessarily be granted access to all information. For example, an individual charged with mailing blank ballots needs access to voter names, addresses and residency information, but may not need access to sensitive voter information used for authentication purposes. Officials should regularly review their access control policies and make appropriate changes as the individuals' responsibilities change.

Identity and Authentication (IA):
Officials should develop technical and procedural mechanisms to identify all users of the election system, including voters, election officials and system administrators. Voter authentication in postal mail systems is largely done via procedural mechanisms. Individual jurisdictions must determine appropriate voter authentication mechanisms. Initial voter

---

[1] Registration and Ballot Return
[2] Ballot Delivery
[3] Ballot Return

authentication occurs in the registration phase, where some type of authenticator (typically a voter signature) is exchanged. This authenticator must be securely stored by election officials so that it is available to authenticate future correspondence from a voter. While some authenticators, such as PINs, are easy to verify, training is necessary to verify authenticators like voter signatures.

Authentication on election IT systems should be automated and tied to the systems' access control and auditing mechanisms. Systems should identify and authenticate each individual with access to a system, usually through a user name and password. Jurisdictions should develop appropriate policies regarding the use of passwords for authentication, including setting password complexity requirements and expiration times, or the use of biometrics.

Incident Response (IR):
Election officials should monitor vital necessary election components to ensure they are functioning properly. Postal mail systems may use a combination of computer and manual systems and procedures. Officials should monitor audit records of electronic voter registration databases and automated ballot tracking systems. Officials should also continuously monitor access to physical storage locations of registration forms and ballots. Also, officials should monitor the status of the postal mail system, watching for current mail disruptions and events which could cause disruptions in the future. While it may be difficult to recover from events in a current election, detected incidents may suggest important technical and procedural controls for future elections.

Media Protection (MP):
Examples of election media in postal mail election systems are registration forms, blank ballots and voted ballots, all of which are on paper. Access to these forms and ballots should be tightly controlled. This media should be stored in a secure location. Only election officials involved with the absentee voting process should have access to this physical location, and any accesses should be logged procedurally or, preferably, automatically.

Officials have limited control of registration forms and ballots in the mail. However, officials should track items, particularly ballots, through the mail whenever possible. A number of deployed absentee ballot management systems exist which provide ballot tracking capabilities. This functionality is not only useful for tracking ballots through the mail, but also throughout the entire voting process, from ballot casting to counting. Such tracking systems can mitigate a large number of internal and external threats to postal mail election systems. However, they also present a privacy risk. Ballot tracking systems should implement procedural and technical controls which can be used to maintain voter privacy.

Physical Security (PE):
As discussed, it is important to limit physical access to election systems and voter information. Physical access to storage locations of registration forms and ballots, inbound and outbound mail boxes and vital election IT systems should be limited to only those who need access to perform their election-related duties. Access could be limited using locks and/or keycards. Whenever possible, access to these locations should be logged.

Personnel Security (PS):
A malicious election official or system administrator could attack a postal mail system in a variety of ways. Jurisdictions should categorize the various roles in their election process according to the level of access to voter information and ballots. Whenever possible, the confidentiality, integrity or availability of the election system should not depend on a single individual. However, that may be infeasible. Some individuals, such as the person charged with addressing and mailing blank ballots, could inflict harm on the system, and it may not be feasible to do all tasks in pairs. In such instances, jurisdictions should do whatever is necessary to gain confidence that that individual will perform his or her duties appropriately. This may include some kind of background screening process.

7.2 Telephone Transmission

Ctrl. Number	Stages[4] RB	BR	Control Text	Notes:
AC-2 X			ACCOUNT MANAGEMENT	IT systems used to manage registration information and interact with voters using telephone lines should be protected with access control mechanisms.
AC-3 X			ACCESS ENFORCEMENT	
AC-5 X			SEPARATION OF DUTIES	
AC-6 X			LEAST PRIVILEGE	
AU-2 X			AUDITABLE EVENTS	Election systems should include auditing functionality to determine the actions of users. This should be done via automated means on IT systems, such as the electronic registration database and telephone voting server, or via procedural methods to record manual actions by election officials.
AU-3 X			CONTENT OF AUDIT RECORDS	
AU-4 X			AUDIT STORAGE CAPACITY	
AU-6 X			AUDIT MONITORING, ANALYSIS, AND REPORTING	
AU-7 X			AUDIT REDUCTION AND REPORT GENERATION	
AU-8 X			TIME STAMPS	
AU-9 X			PROTECTION OF AUDIT INFORMATION	
AU-10 X			NON-REPUDIATION	
CM-2 X			BASELINE CONFIGURATION	System administrators should closely monitor the configuration of vital IT systems to ensure they have not been manipulated.
CM-3 X			CONFIGURATION CHANGE CONTROL	
CM-5 X			ACCESS RESTRICTIONS FOR CHANGE	
CP-7 X		X	ALTERNATE PROCESSING SITE	Officials should prepare a backup telecommunications system in case of an unscheduled outage or attack.
CP-8 X		X	TELECOMMUNICATIONS SERVICES	
IA-1 X		X	IDENTIFICATION AND AUTHENTICATION POLICY AND PROCEDURES	Officials should develop procedures and implement technical mechanisms to identify voters, election officials and system administrators. Procedures may be used to authenticate voters, while IT systems should include IA and access control mechanisms.
IA-2 X		X	USER IDENTIFICATION AND AUTHENTICATION	
IA-4 X		X	IDENTIFIER MANAGEMENT	
IA-5 X		X	AUTHENTICATOR MANAGEMENT	
IA-7 X		X	CRYPTOGRAPHIC MODULE AUTHENTICATION	
IR-4 X		X	INCIDENT HANDLING	Officials should monitor their IT systems and communications services for disruptions and possible attacks.
IR-5 X		X	INCIDENT MONITORING	
MA-2 X			CONTROLLED MAINTENANCE	System administrators should closely monitor the maintenance of vital IT systems to ensure the proper hardware and software updates are performed on such systems.
MA-3 X			MAINTENANCE TOOLS	
MA-5 X			MAINTENANCE PERSONNEL	
PE-2 X		X	PHYSICAL ACCESS AUTHORIZATIONS	Officials should control physical access to vital systems and sensitive information. This includes physical access to IT systems and communications equipment.
PE-3 X		X	PHYSICAL ACCESS CONTROL	
PE-4 X		X	ACCESS CONTROL FOR TRANSMISSION MEDIUM	
PE-5 X		X	ACCESS CONTROL FOR DISPLAY MEDIUM	
PE-6 X		X	MONITORING PHYSICAL ACCESS	

---

[4] Telephones are not used to create a ballot delivery system. Telephone voting systems provide voters with ballot questions, along with a mechanism to submit votes. As such, telephone voting systems are considered a type of ballot return system.

Ctrl. Number	Stages[4]		Control Text	Notes:
	RB	BR		
PS-2	X	X	POSITION CATEGORIZATION	Screen election employees who will be administering IT systems or receiving/transcribing voter information.
PS-3	X	X	PERSONNEL SCREENING	
SC-5	X	X	DENIAL OF SERVICE PROTECTION	Officials must develop protections against denial of service attacks and mitigate the effect with backup procedures.
SC-8	X	X	TRANSMISSION INTEGRITY	
SC-9	X	X	TRANSMISSION CONFIDENTIALITY	Whenever possible, information sent over telephone lines should have integrity and confidentiality protection. This may be possible with kiosks with secure telephones.
SC-12	X	X	CRYPTOGRAPHIC KEY ESTABLISHMENT AND MANAGEMENT	
SC-13	X	X	USE OF CRYPTOGRAPHY	
SI-2	X		FLAW REMEDIATION	System administrators should watch for defects and malicious code in IT systems that could prevent those systems from functioning properly.
SI-3	X		MALICIOUS CODE PROTECTION	
SI-4	X		INFORMATION SYSTEM MONITORING TOOLS AND TECHNIQUES	
SI-7	X		SOFTWARE AND INFORMATION INTEGRITY	

Access Control (AC):
Most telephone-based voting systems contain at least two vital computer systems: an electronic voter registration database and the telephone voting server, which connects to the telephone network and interacts with voters. Both of these systems need to be protected by access control mechanisms. Access to these systems should be limited to employees who need the information on the system to perform election-related duties. If election officials are communicating directly with voters registering or requesting ballots, several officials may need access to the database. Fewer individuals should need access to the information on the telephone voting server, as it automates the process of interacting with voters. Officials should regularly review their access control policies and make appropriate changes as individuals' responsibilities change.

Audit and Accountability (AU):
Election computer systems should keep audit records of important events on the system, such as authentication attempts, maintenance and other administrative activities, and voter sessions. The audit records should provide enough information to determine who performed a given action, a description of the action, and the time it took place.

Maintaining the integrity of this information is important, and systems should implement controls which protect audit information from unauthorized access, modification or deletion. The security control AU-9(1) listed in NIST SP-800-53 suggests that audit records be produced on hardware-enforced, write-once media. This control, or variations of it, is highly recommended for important election records, such as votes. Systems could print certain kinds of election records on paper, and store them in a secure box. Alternatively, systems could implement cryptographic protections, such as signing records using validated hardware cryptographic modules validated under FIPS-140, Security Requirements for Cryptographic Modules, procedures.

Configuration Management (CM):
The integrity of votes in a telephone voting system is dependent on the software in the telephone voting server. System administrators should have a baseline configuration for the election system, and access control mechanisms should prevent anyone other than authorized system administrators from making any changes to this configuration. All changes should be recorded in the audit log for the system.

Contingency Planning (CP):
Backup plans and systems should be developed and implemented in the event that telephone service drops due to increased demand, outages, or attacks.

Identity and Authentication (IA):
Officials should develop technical and procedural mechanisms to identify all users of the election system, including voters, election officials and system administrators. Voter authentication in registration systems is largely done via procedural mechanisms. Individual jurisdictions must determine appropriate voter authentication mechanisms. Initial voter authentication occurs in the registration phase, where some type of authenticator is exchanged. In the case of telephone voting systems, the voter authenticator is likely a PIN. This authenticator must be securely stored by election officials so that it is available to authenticate future correspondence from a voter.

Authentication on election IT systems should be automated and tied to the systems' access control and auditing mechanisms. Systems should identify and authenticate each individual with access to a system, usually through a user name and password. Jurisdictions should develop appropriate policies regarding the use of passwords for authentication, including setting password complexity requirements and expiration times.

Incident Response (IR):
Election officials should monitor vital necessary election systems and communications services to ensure they are functioning properly. Officials should monitor audit records of electronic voter registration databases and telephone voting system servers. Officials should also continuously monitor physical access to these systems. To protect against unscheduled service outages and denial of service attacks, administrators should closely monitor the status of the telephone lines used to register voters and submit votes, and implement contingency plans when necessary.

Maintenance (MA):
Telephone voting systems rely on software to ensure that votes are recorded properly. Due to potential software defects, it is important for jurisdictions to develop and follow appropriate controls to see that software updates are installed when needed. Because of the threat of malicious code, it is important that these controls ensure that only proper software updates are installed, and that these updates are installed by authorized system administrators.

Physical Security (PE):
It is important to limit physical access to election systems and voter information. Individuals with physical access to election computer systems may be able to access sensitive records, modify records or software, or cause equipment to fail. Access to areas containing vital election

systems should be limited to only those who need access to perform their election-related duties. Access could be limited using locks and/or keycards. Whenever possible, access to these locations should be logged.

Personnel Security (PS):
A malicious election official or system administrator may have access to vital election system equipment or information. Jurisdictions should categorize the various roles in their election process according to the level of access to voter information and ballots. Whenever possible, the confidentiality, integrity or availability of the election system should not depend on a single individual. However, that may not be feasible. One jurisdiction may not have multiple employees capable of acting as system administrators for the electronic registration database or the telephone voting server. Jurisdictions should take appropriate actions to gain confidence that that individual will perform his or her duties appropriately. This may include some kind of background screening process.

System and Communications Protection (SC):
Sensitive or critical information transmitted over a public communications network typically should have cryptographic protections in order to protect the confidentiality and/or integrity of transmitted data. However, such protections could not be implemented without preventing voters with standard telephones from using the telephone voting system. An alternative is for jurisdictions to set up kiosks with secure telephones. Voters would not be able to vote from their home telephones; instead they would have to go to a kiosk and vote from one of the terminals. Individual jurisdictions must weigh the risks of eavesdropping and modifications in transit against the convenience of telephone voting from home.

System and Information Integrity (SI):
As previously noted, telephone voting systems rely on the correctness of software running on the telephone voting server. System administrators should test and monitor their systems to look for defects in the system that could prevent votes from being recorded properly, disrupt the elections, or release sensitive information to an attacker. Furthermore, election computer systems should be protected from malicious code using antivirus software. Systems connected to a network should be protected with a firewall and an intrusion detection system (IDS). In most cases the firewall and IDS will be separate devices on the jurisdiction's computer network.

7.3 Fax Transmission

Ctrl. Number	Stages RB	BD	BR	Control Text	Notes
AC-2	X			ACCOUNT MANAGEMENT	Databases and IT systems used to manage registration information should be protected with access control mechanisms.
AC-3	X			ACCESS ENFORCEMENT	
AC-5	X			SEPARATION OF DUTIES	
AC-6	X			LEAST PRIVILEGE	
CP-7	X	X		ALTERNATE PROCESSING SITE	Officials should prepare a backup telecommunications system in case of an unscheduled outage or attack.
CP-8	X	X		TELECOMMUNICATIONS SERVICES	
IA-1	X	X		IDENTIFICATION AND AUTHENTICATION POLICY AND PROCEDURES	Officials should develop procedures and implement technical mechanisms to identify voters, election officials and system administrators. Procedures may be used to authenticate voters, while IT systems should include IA and access control mechanisms.
IA-2	X	X		USER IDENTIFICATION AND AUTHENTICATION	
IA-4	X	X		IDENTIFIER MANAGEMENT	
IA-5	X	X		AUTHENTICATOR MANAGEMENT	
IA-7	X	X		CRYPTOGRAPHIC MODULE AUTHENTICATION	
IR-4	X	X	X	INCIDENT HANDLING	Officials should monitor their IT systems and communications services for disruptions and possible attacks.
IR-5	X	X	X	INCIDENT MONITORING	
MP-1	X	X	X	MEDIA PROTECTION POLICY AND PROCEDURES	Officials should protect registration forms, blank paper ballots and voted ballots with procedures.
MP-2	X	X	X	MEDIA ACCESS	
MP-4	X	X	X	MEDIA STORAGE	
PE-2	X	X	X	PHYSICAL ACCESS AUTHORIZATIONS	Officials should control physical access to vital systems and sensitive information. This includes physical access to IT systems and communications equipment.
PE-3	X	X	X	PHYSICAL ACCESS CONTROL	
PE-4	X			ACCESS CONTROL FOR TRANSMISSION MEDIUM	
PE-6	X	X		MONITORING PHYSICAL ACCESS	
PS-2	X	X	X	POSITION CATEGORIZATION	Screen election employees who will be handling registration forms and ballots.
PS-3	X	X	X	PERSONNEL SCREENING	
SC-5	X	X	X	DENIAL OF SERVICE PROTECTION	Officials must develop protections against denial of service attacks or mitigate the effect with backup procedures.
SC-8	X	X	X	TRANSMISSION INTEGRITY	
SC-9	X	X	X	TRANSMISSION CONFIDENTIALITY	Whenever possible, information sent over telephone lines should have integrity and confidentiality protection. This may be possible with kiosks holding secure fax machines.
SC-12	X	X	X	CRYPTOGRAPHIC KEY ESTABLISHMENT AND MANAGEMENT	
SC-13	X	X	X	USE OF CRYPTOGRAPHY	
SC-14	X			PUBLIC ACCESS PROTECTIONS	

Access Control (AC):
IT systems containing important election information, such as an electronic voter registration database should be protected by access control mechanisms. Access to these systems should be limited to employees who need this information to perform election-related duties. Furthermore, individuals who need access to some voter information should not necessarily be granted access to all information. Officials should regularly review their access control policies and make appropriate changes as individuals' responsibilities change.

Contingency Planning (CP):
Backup plans and systems should be developed and implemented in the event that telephone service drops due to increased demand, outages, or attacks.

Identity and Authentication (IA):
Officials should develop technical and procedural mechanisms to identify all users of the election system, including voters, election officials and system administrators. Voter authentication in fax systems is largely done via procedural mechanisms. Individual jurisdictions must determine appropriate voter authentication mechanisms. Initial voter authentication occurs in the registration phase, where some type of authenticator (typically a voter signature) is exchanged. This authenticator must be securely stored by election officials so that it is available to authenticate future correspondence from a voter. While some authenticators, such as PINs, are easy to verify, training is necessary to verify authenticators like voter signatures.

Authentication on election IT systems should be automated and tied to the systems' access control and auditing mechanisms. Systems should identify and authenticate each individual with access to a system, usually through a user name and password. Jurisdictions should develop appropriate policies regarding the use of passwords for authentication, including setting password complexity requirements and expiration times.

Incident Response (IR):
Election officials should monitor vital necessary election systems, such as the voter registration database, and communications services to ensure they are functioning properly. Officials should also continuously monitor physical access to these systems. To protect against unscheduled service outages and denial of service attacks, administrators should closely monitor the status of the telephone lines used to receive faxed requests and ballots, and implement contingency plans when necessary.

Media Protection (MP):
Examples of election media in election systems are registration forms, blank ballots and voted ballots, all of which are on paper prior to and after being faxed. Access to these forms and ballots should be tightly controlled. This media should be stored in a secure location. Only election officials involved with the absentee voting process should have access to this physical location, and any accesses should be logged procedurally or, preferably, automatically. Specifically, registration forms and voted ballots received via fax are at-risk to being read or modified by anyone in the vicinity of the fax machine. Fax machines that will receive election materials should be kept in a locked room.

Election materials in fax-based systems will experience two or more conversions from being a physical entity to an electronic signal, or vice versa. While paper-based systems can use unique ballot stock to help identify clearly forged ballots, this is not possible in a fax-based system. Attackers could make multiple copies of ballots, using them to flood election official offices or perform other attacks. Ballot tracking systems, such as those described Section 6.1, could help mitigate this threat, while also helping ensure paper copies of faxed ballots are not lost in the counting process. The systems used with postal ballots should be able to be used with minor modifications.

Physical Security (PE):
It is important to limit physical access to election systems and voter information. As previously noted, individuals with physical access to fax machines may be able to read sensitive voter information, violate voter privacy or modify received votes. Access to areas containing vital election systems, including fax machines and voter registration databases, should be limited only to those who need access to perform their election-related duties. Access could be limited using locks and/or keycards. Whenever possible, access to these locations should be logged.

Personnel Security (PS):
As previously discussed, a malicious election official or system administrator could attack a fax-based election system in a variety of ways. Jurisdictions should categorize the various roles in their election process according to the level of access to voter information and ballots. Whenever possible, the confidentiality, integrity or availability of the election system should not depend on a single individual. However, that may not be feasible. Some individuals, such as the person charged with faxing blank ballots, could inflict harm on the system, and it may not be feasible to do all tasks in pairs. In such instances, jurisdictions should do whatever is necessary to gain confidence that that individual will perform his or her duties appropriately. This may include some kind of background screening process.

System and Communications Protection (SC):
Sensitive or critical information transmitted over a public communications network may have cryptographic protections in order to protect the confidentiality and/or integrity of transmitted data. However, such protections could not be implemented without preventing voters with a standard fax machine from using the system. An alternative is for jurisdictions to set up kiosks with secure fax machines. Voters would not be able to vote from their home fax machines; instead they would have to go to a kiosk and vote from one of the terminals. Individual jurisdictions must weigh the risks of eavesdropping and modifications in transit against the convenience of voting using a fax machine at home.

7.4 E-Mail Transmission

Ctrl. No.	Stages			Control Name	Notes
	RB	BD	BR		
AC-2	X	X	X	ACCOUNT MANAGEMENT	IT systems used to manage registration information, election workstations, and local e-mail servers should be protected with access control mechanisms.
AC-3	X	X	X	ACCESS ENFORCEMENT	
AC-5	X	X	X	SEPARATION OF DUTIES	
AC-6	X	X	X	LEAST PRIVILEGE	
AC-12	X			SESSION TERMINATION	
AU-2	X			AUDITABLE EVENTS	Election systems should include auditing functionality to determine. E-mail servers, election workstations, and registration databases should have system event logging functionality. Procedural methods should be used to record manual actions by election officials.
AU-3	X			CONTENT OF AUDIT RECORDS	
AU-4	X			AUDIT STORAGE CAPACITY	
AU-6	X			AUDIT MONITORING, ANALYSIS, AND REPORTING	
AU-7	X			AUDIT REDUCTION AND REPORT GENERATION	
AU-8	X			TIME STAMPS	
AU-9	X			PROTECTION OF AUDIT INFORMATION	
AU-10	X			NON-REPUDIATION	
CM-2	X			BASELINE CONFIGURATION	System administrators should closely monitor the maintenance of vital IT systems to ensure the proper hardware and software updates are performed on such systems.
CM-3	X			CONFIGURATION CHANGE CONTROL	
CM-5	X			ACCESS RESTRICTIONS FOR CHANGE	
CP-7	X	X	X	ALTERNATE PROCESSING SITE	Officials should prepare a backup telecommunications system in case of an unscheduled outage or attack.
CP-8	X	X	X	TELECOMMUNICATIONS SERVICES	
IA-1	X	X		IDENTIFICATION AND AUTHENTICATION POLICY AND PROCEDURES	Officials should develop procedures and implement technical mechanisms to identify voters, election officials and system administrators. Procedures may be used to authenticate voters, while IT systems should include IA and access control mechanisms.
IA-2	X	X		USER IDENTIFICATION AND AUTHENTICATION	
IA-4	X	X		IDENTIFIER MANAGEMENT	
IA-5	X	X		AUTHENTICATOR MANAGEMENT	
IA-7	X	X		CRYPTOGRAPHIC MODULE AUTHENTICATION	
IR-4	X	X	X	INCIDENT HANDLING	Officials should monitor their IT systems and communications services for disruptions and possible attacks.
IR-5	X	X	X	INCIDENT MONITORING	
MA-2	X			CONTROLLED MAINTENANCE	System administrators should closely monitor the maintenance of vital IT systems to ensure the proper hardware and software updates are performed on such systems.
MA-3	X			MAINTENANCE TOOLS	
MA-5	X			MAINTENANCE PERSONNEL	
MP-1	X			MEDIA PROTECTION POLICY AND PROCEDURES	Officials should protect printed returned ballots with procedures.
MP-2	X			MEDIA ACCESS	
MP-4	X			MEDIA STORAGE	
PE-2	X	X	X	PHYSICAL ACCESS AUTHORIZATIONS	Officials should control physical access to vital systems and sensitive information. This includes physical access to IT systems and communications equipment.
PE-3	X	X	X	PHYSICAL ACCESS CONTROL	
PE-4	X	X		ACCESS CONTROL FOR TRANSMISSION MEDIUM	
PE-6	X			MONITORING PHYSICAL ACCESS	

Ctrl. No.	Stages RB	BD	BR	Control Name	Notes
PS-2	X	X	X	POSITION CATEGORIZATION	Screen election employees who will be administering IT systems.
PS-3	X	X	X	PERSONNEL SCREENING	
SC-5	X	X		DENIAL OF SERVICE PROTECTION	Officials must develop protections against denial of service attacks or mitigate the effect with backup procedures. Information sent over e-mail should have integrity protection and, if possible, confidentiality protection.
SC-7	X	X	X	BOUNDARY PROTECTION	
SC-8	X	X		TRANSMISSION INTEGRITY	
SC-9	X	X		TRANSMISSION CONFIDENTIALITY	
SC-12	X		X	CRYPTOGRAPHIC KEY ESTABLISHMENT AND MANAGEMENT	
SC-13	X	X	X	USE OF CRYPTOGRAPHY	
SC-14	X	X		PUBLIC ACCESS PROTECTIONS	
SC-20	X	X	X	SECURE NAME / ADDRESS RESOLUTION SERVICE (AUTHORITATIVE SOURCE)	
SC-21	X	X	X	SECURE NAME / ADDRESS RESOLUTION SERVICE (RECURSIVE OR CACHING RESOLVER)	
SI-2	X	X		FLAW REMEDIATION	System administrators should watch for defects and malicious code in IT systems that could prevent those systems from functioning properly.
SI-3	X	X		MALICIOUS CODE PROTECTION	
SI-4	X	X		INFORMATION SYSTEM MONITORING TOOLS AND TECHNIQUES	
SI-5	X	X		SECURITY ALERTS AND ADVISORIES	
SI-7	X	X		SOFTWARE AND INFORMATION INTEGRITY	

Access Control (AC):
E-mail-based voting systems contain several different computer systems vital to the election process. These include computer workstations used by election officials, voter registration databases, and local e-mail servers administered by the jurisdiction. All of these systems should be protected by access control mechanisms. Access to these systems should be limited to employees who need this information to perform election-related duties. Officials should regularly review their access control policies and make appropriate changes as individuals' responsibilities change.

Audit and Accountability (AU):
Election computer systems should keep audit records of important events on the system, such as authentication attempts, maintenance and other administrative activities, and voter sessions. The audit records should provide enough information to determine who performed a given action, a description of the action, and the time it took place.

Maintaining the integrity of this information is important, and systems should implement controls which protect audit information from unauthorized access, modification or deletion. The security control AU-9(1) listesd in NIST SP-800-53 suggests that audit records be produced on hardware-enforced, write-once media. This control, or variations of it, is highly recommended for important election records, such as votes. Election officials should consider printing received ballots immediately, and storing them in a secure location. It should be noted, however, that this would merely duplicate the results of many attacks, rather than prevent them. For instance, if voted ballots are modified before reaching the election official, printing the modified ballots would not prevent or detect the attack.

Configuration Management (CM):
The integrity of votes and the reliability of the system are dependent on the correctness of software in key computer systems supporting the election process. Computer workstations, voter registration databases and e-mail servers are all vital election computer systems. System administrators should have baseline configurations for these systems, and access control mechanisms should prevent anyone other than authorized system administrators from making any changes to these configurations. All changes should be recorded in the audit log for the system.

Contingency Planning (CP):
Backup plans and systems should be developed and implemented in the event that Internet service drops due to increased demand, outages, or attacks.

Identity and Authentication (IA):
Officials should develop technical and procedural mechanisms to identify all users of the election system, including voters, election officials and system administrators. Voter authentication in registration systems is largely done via procedural mechanisms. Individual jurisdictions must determine appropriate voter authentication mechanisms. Voters must provide election officials with an authenticator during the registration phase. For election systems using e-mail ballot return, the most likely authenticator is a voter signature. Authenticators must be securely stored by election officials so that it is available to authenticate future correspondence from a voter. Other systems may use passwords, PINs, or digital signatures.

Authentication on election IT systems should be automated and tied to the systems' access control and auditing mechanisms. Systems should identify and authenticate each individual with access to a system, usually through a user name and password. Jurisdictions should develop appropriate policies regarding the use of passwords for authentication, including setting password complexity requirements and expiration times.

Incident Response (IR):
Election officials should monitor vital election systems and communications services to ensure they are functioning properly. Officials should monitor audit records of electronic voter registration databases and e-mail servers to verify they are functioning correctly. Officials should also continuously monitor physical access to these systems. To protect against unscheduled service outages and denial of service attacks, administrators should closely monitor the status of Internet connections, and the storage space available on e-mail servers, and implement contingency plans when necessary.

Maintenance (MA):
E-mail-based voting systems rely on the software on e-mail servers to assure that election integrity is maintained and that systems remain available to the public. Due to potential software defects in these systems, and the fact that they are connected to the Internet, it is important for

jurisdictions to develop and follow appropriate controls to see that software updates are installed when needed. Because of the threat of malicious code, it is important that these controls ensure that only proper software updates are installed, and that these updates are installed by authorized system administrators.

Physical Security (PE):
It is important to limit physical access to election computer systems and voter information. Individuals with physical access to election computer systems may be able to access sensitive records, modify records or software, or cause equipment to fail. Furthermore, individuals with access to storage locations for printed ballots may be able to violate voter privacy or modify votes. Access to areas containing vital election systems should be limited to only those who need access to perform their election-related duties. Access could be limited using locks and/or keycards. Whenever possible, access to these locations should be logged.

Personnel Security (PS):
A malicious election official or system administrator may have access to vital election system equipment or information. Jurisdictions should categorize the various roles in their election process according to the level of access to voter information and ballots. Whenever possible, the confidentiality, integrity or availability of the election system should not depend on a single individual. However, that may be infeasible. One jurisdiction may not have multiple employees capable of acting as system administrators for election computer systems. Jurisdictions should take appropriate actions to gain confidence that the administrator will perform his or her duties appropriately. This may include some kind of background screening process.

System and Communications Protection (SC):
Sensitive or critical information transmitted over a public communications network typically should have cryptographic protections in order to protect the confidentiality and/or integrity of transmitted data. By itself, e-mail offers little support for cryptographic functionality. However, e-mail based election systems mainly use e-mail to transfer files, such as registration forms or ballots. These files could be cryptographically protected.

Election officials should digitally sign all registration forms and blank ballots before distributing them to voters through e-mail. The Portable Document Format (PDF) files can be digitally signed in some applications that create them. With the correct software, voters' computers will automatically check the digital signature and warn voters of any problems. Such a system would require election officials to create a Digital Signature Standard (DSS) or RSA key pair [23] and apply for a digital certificate from a major certificate vendor. However, only election officials would need to obtain a key pair and certificate.

However, at this time there is no practical way to protect the integrity or confidentiality of e-mails from voters. Thus, returned ballots would be at risk for eavesdropping and modification. Individual jurisdictions must weigh these risks against the convenience of returning ballots via e-mail. S/MIME [22] is a possible solution for digitally signing and encrypting e-mails if voters and elections are able to obtain key pairs and digital certificates. This would require the deployment of a large-scale Public Key Infrastructure.

System and Information Integrity (SI):
As previously noted, e-mail voting systems rely on the software running on election computer systems such as e-mail servers and workstations. System administrators should test and monitor their systems to look for defects or vulnerabilities that could prevent votes from being recorded properly, disrupt the elections, or release sensitive information to an attacker. Administrators can check sources, such as the National Vulnerability Database [26], for new security problems with their systems. Furthermore, election computer systems should be protected from various software and network attacks using antivirus software, firewalls and intrusion detection systems.

7.5 Web-Based Transmission

Ctrl. No.	RB	BD	BR	Control Name	Notes
AC-2	X	X	X	ACCOUNT MANAGEMENT	IT systems used to manage registration information and record votes should be protected with access control mechanisms.
AC-3	X	X	X	ACCESS ENFORCEMENT	
AC-5	X	X	X	SEPARATION OF DUTIES	
AC-6	X	X	X	LEAST PRIVILEGE	
AC-12	X			SESSION TERMINATION	
AU-2	X			AUDITABLE EVENTS	Election systems should include auditing functionality to determine. All computer systems involved in the election process should include system event log functionality. Procedural methods should be used to record manual actions by election officials.
AU-3	X			CONTENT OF AUDIT RECORDS	
AU-4	X			AUDIT STORAGE CAPACITY	
AU-6	X			AUDIT MONITORING, ANALYSIS, AND REPORTING	
AU-7	X			AUDIT REDUCTION AND REPORT GENERATION	
AU-8	X			TIME STAMPS	
AU-9	X			PROTECTION OF AUDIT INFORMATION	
AU-10	X			NON-REPUDIATION	
CM-2	X	X		BASELINE CONFIGURATION	System administrators should closely monitor the maintenance of vital IT systems to ensure the proper hardware and software updates are performed on such systems.
CM-3	X	X		CONFIGURATION CHANGE CONTROL	
CM-5	X	X		ACCESS RESTRICTIONS FOR CHANGE	
CP-7	X	X	X	ALTERNATE PROCESSING SITE	Officials should prepare a backup telecommunications system in case of an unscheduled outage or attack.
CP-8	X	X	X	TELECOMMUNICATIONS SERVICES	
IA-1	X	X		IDENTIFICATION AND AUTHENTICATION POLICY AND PROCEDURES	Officials should develop procedures and implement technical mechanisms to identify voters, election officials and system administrators. Procedures may be used to authenticate voters, while IT systems should include authentication and access control mechanisms.
IA-2	X	X		USER IDENTIFICATION AND AUTHENTICATION	
IA-4	X	X		IDENTIFIER MANAGEMENT	
IA-5	X	X		AUTHENTICATOR MANAGEMENT	
IA-7	X	X		CRYPTOGRAPHIC MODULE AUTHENTICATION	
IR-4	X	X	X	INCIDENT HANDLING	Officials should monitor their IT systems and communications services for disruptions and possible attacks.
IR-5	X	X	X	INCIDENT MONITORING	
MA-2	X	X		CONTROLLED MAINTENANCE	System administrators should closely monitor the maintenance of vital IT systems to ensure the proper hardware and software updates are performed on such systems.
MA-3	X	X		MAINTENANCE TOOLS	
MA-5	X	X		MAINTENANCE PERSONNEL	
PE-2	X	X	X	PHYSICAL ACCESS AUTHORIZATIONS	Officials should control physical access to vital systems and sensitive information. This includes physical access to IT systems and communications equipment.
PE-3	X	X	X	PHYSICAL ACCESS CONTROL	
PE-4	X	X		ACCESS CONTROL FOR TRANSMISSION MEDIUM	
PE-5	X			ACCESS CONTROL FOR DISPLAY MEDIUM	
PE-6	X	X		MONITORING PHYSICAL ACCESS	
PS-2	X	X	X	POSITION CATEGORIZATION	Screen election employees who will be administering IT systems.
PS-3	X	X	X	PERSONNEL SCREENING	

Ctrl. No.	Stages			Control Name	Notes
	RB	BD	BR		
SC-5	X	X		DENIAL OF SERVICE PROTECTION	Officials must develop protections against denial of service attacks or mitigate the effect with backup procedures. Information sent over telecommunication lines should have integrity and confidentiality protection.
SC-7	X	X	X	BOUNDARY PROTECTION	
SC-8	X	X		TRANSMISSION INTEGRITY	
SC-9	X	X		TRANSMISSION CONFIDENTIALITY	
SC-12	X		X	CRYPTOGRAPHIC KEY ESTABLISHMENT AND MANAGEMENT	
SC-13	X	X	X	USE OF CRYPTOGRAPHY	
SC-14	X	X		PUBLIC ACCESS PROTECTIONS	
SC-20	X	X	X	SECURE NAME / ADDRESS RESOLUTION SERVICE (AUTHORITATIVE SOURCE)	
SC-21	X	X	X	SECURE NAME / ADDRESS RESOLUTION SERVICE (RECURSIVE OR CACHING RESOLVER)	
SI-2	X	X		FLAW REMEDIATION	System administrators should watch for defects and malicious code in IT systems that could prevent those systems from functioning properly
SI-3	X	X		MALICIOUS CODE PROTECTION	
SI-4	X	X		INFORMATION SYSTEM MONITORING TOOLS AND TECHNIQUES	
SI-5	X	X		SECURITY ALERTS AND ADVISORIES	
SI-7	X	X		SECURITY ALERTS AND ADVISORIES	

Access Control (AC):
Web-based registration and voting systems rely on a web server to interact with voters and store vital election information. This system must be protected by strict access control mechanisms. It is likely that some information may be moved to other computer systems. For instance, registration information may be moved to a voter registration database, and tallied votes may be moved to an election management system. Access to information stored on these devices should be limited to employees who need this information to perform election-related duties. Officials should regularly review their access control policies and make appropriate changes as individuals' responsibilities change.

Audit and Accountability (AU):
Election computer systems should keep audit records of important events on the system, such as authentication attempts, maintenance and other administrative activities, and voter sessions. The audit records should provide enough information to determine who performed a given action, a description of the action, and the time it took place.

Maintaining the integrity of this information is important, and systems should implement controls which protect audit information from unauthorized access, modification or deletion. Audit records on vital election systems should be digitally signed, preferably using a hardware cryptographic module.

Configuration Management (CM):
The integrity of votes and the reliability of the system are dependent on the correctness of software in election web server. System administrators should have a baseline configuration for this system, and access control mechanisms should prevent anyone other than authorized system

administrators from making any changes to this configuration. All changes should be recorded in the audit log for the system.

Contingency Planning (CP):
Backup plans and systems should be developed and implemented in the event that Internet service drops due to increased demand, outages, or attacks.

Identity and Authentication (IA):
Officials should develop technical and procedural mechanisms to identify all users of the election system, including voters, election officials and system administrators. Initial voter authentication is often done via procedural means. Online voter authentication may need to be done using secret information from the voter. In all cases, each voter must share an authenticator with election officials during the registration phase. Typical authenticators for online systems include passwords, PINs or digital certificates. Authenticators must be securely stored by election officials so that they are available to authenticate future correspondence from a voter.

Authentication on election IT systems should be automated and tied to the systems' access control and auditing mechanisms. Systems should identify and authenticate each individual with access to a system, usually through a user name and password. Jurisdictions should develop appropriate policies regarding the use of passwords for authentication, including setting password complexity requirements and expiration times.

Incident Response (IR):
Election officials should monitor vital necessary election systems and communications services to ensure they are functioning properly. Officials should monitor audit records of electronic voter registration databases and election web servers to verify that they are functioning correctly. Officials should also continuously monitor physical access to these systems. To protect against unscheduled service outages and denial of service attacks, administrators should closely monitor the status of Internet connections and implement contingency plans when necessary.

Maintenance (MA):
Web-based voting systems rely on the correctness of software to ensure election integrity and availability. Defects and vulnerabilities in voting system software could violate the security goals of the system. System administrators should watch for new vulnerabilities in their systems by monitoring sites such as the National Vulnerability Database [26], and check for updates from software manufacturers. It is important for jurisdictions to develop and follow appropriate controls to see that software updates are installed when needed. Because of the threat of malicious code, it is important that these controls ensure that only proper software updates are installed, and that these updates are installed by authorized system administrators.

Physical Security (PE):
It is important to limit physical access to election computer systems and voter information. Individuals with physical access to election computer systems may be able to access sensitive records, modify records or software, or cause equipment to fail. Access to areas containing vital election systems should be limited to only those who need access to perform their election-

related duties. Access could be limited using locks and/or keycards. Whenever possible, access to these locations should be logged.

Personnel Security (PS):
A malicious election official or system administrator may have access to vital election system equipment or information. Jurisdictions should categorize the various roles in their election process according to the level of access to voter information and ballots. Whenever possible, the confidentiality, integrity or availability of the election system should not depend on a single individual. However, that may be infeasible. One jurisdiction may not have multiple employees capable of acting as system administrators for election computer systems. Jurisdictions should take appropriate actions to gain confidence that that individual will perform his or her duties appropriately. This may include some kind of background screening process.

System and Communications Protection (SC):
Sensitive or critical information transmitted over a public communications network should have cryptographic protections in order to protect the confidentiality and integrity of transmitted data. Web-based election systems should use SSL/TLS to create a secure communications channel between the voter and election web server. Web servers should have a valid SSL certificate from a major certificate vendor. This will allow voters to authenticate the election web server.

For added protection in ballot distribution, election officials should digitally sign all registration forms and blank ballots before posting them on election websites or distributing them to voters online. The Portable Document Format (PDF) files can be digitally signed in some applications that create them. With the correct software, voters' computers will automatically check the digital signature and warn voters of any problems. Such a system would require election officials to create a Digital Signature Standard (DSS) or RSA key pair [23] and apply for a certificate from a major certificate vendor.

System and Information Integrity (SI):
As previously noted, web-based election systems rely on the software running on the election web server. This is particularly true for systems which allow for web-based voting. System administrators should test and monitor their systems to look for defects or vulnerabilities in the system that could prevent votes from being recorded properly, disrupt the elections, or release sensitive information to an attacker. Administrators can check sources, such as the National Vulnerability Database, for new security problems with their systems. Furthermore, election computer systems should be protected from various software and network attacks using antivirus software, firewalls and intrusion detection systems.

8 Conclusions

This paper discusses the current UOCAVA voting process and provides descriptions of the types of voting materials being exchanged between voters and election officials and the various electronic transmission options available. In addition, this paper describes various threats to those different transmission options and what sorts of security-related controls could be employed to counteract the threats. This section draws upon these threats and controls to arrive at initial conclusions regarding use of these transmissions options with registration and blank ballot requests, delivery of blank ballots, and return of voted ballots. This section also identifies potential next steps; areas of research to pursue in further assisting UOCAVA voters.

8.1 Registration and Blank Ballot Request

As noted, all states use the Federal Post Card Application (FPCA) to register military and civilian overseas citizens to register and request ballots. All four transmission options could be used to submit the information required in the FPCA electronically, but use of e-mail and the web present greater challenges at this time.

Use of Telephone and Fax for Registration and Blank Ballot Requests:
Use of telephone systems by UOCAVA voters to transmit registration and blank ballot requests is similar to use of fax machines in that both systems use the same telephone network infrastructure and therefore share many of the same threats. Many of the threats can be mitigated procedurally. For telephone-based systems, however, certain procedural changes would need to be made in authenticating registration and ballot request information, i.e., a voter would have to prove his or her identity over the phone based on information other than a signature on a fax or postal mail form. If election officials can suitably authenticate voters over the telephone or using fax, these technologies could be used to significantly reduce the delivery times needed to send registration and ballot requests.

Use of E-Mail and Web for Registration and Blank Ballot Requests:
E-mail and web options for transmitting registration and blank ballot requests currently pose more challenges than for telephone and fax. Network-based attacks could disrupt communications between voters and election officials, or put sensitive personal information from voters at risk of being intercepted. Less sophisticated attacks, such as the spoofing and phishing common in the banking industry, could trick voters into providing their personal information to attackers. These threats are very similar to those faced by many e-commerce applications. Successful use will depend on using similar best practices and techniques as those developed for e-commerce and other internet applications.

8.2 Delivery of Blank Ballots

In general, the threats affecting delivery of blank ballots to UOCAVA voters pose less serious challenges than the threats for the return of voted ballots; all four transmission options could be used given careful implementation, including technical and procedural controls.

Use of Fax for Delivery of Blank Ballots:
Most threats to faxed delivery of blank ballots can be mitigated procedurally. The remaining threats are both difficult to enact on a large scale and would have a limited effect on the integrity of the election. Faxed delivery of blank ballots could significantly reduce the delivery times compared to postal mail using technology widely deployed today.

Use of E-mail for Delivery of Blank Ballots:
E-mail is widely deployed and could significantly reduce delivery times for a large number of voters. However, e-mail delivery of blank ballots relies on various systems that are not under the control of election officials; network-based attacks could interfere with ballots being received properly. E-mail can be read or modified while in transit and can be easily spoofed such that recipients may believe the received ballot is legitimate when it is not. Technical controls, such as digitally signing ballots, can mitigate some of these threats.

Use of Web for Delivery of Blank Ballots:
As with e-mail, web-based delivery of blank ballots also relies on various systems that are not under the control of election officials. Network-based attacks pose some threat to these systems, although most can be effectively mitigated using technical controls. Web-based delivery of blank ballots offers some advantages over e-mail in that communications with web servers are more readily protected using widely available security features built into most browsers. While it is difficult to prevent less sophisticated attacks, such as spoofing, the web would offer a convenient and quick ballot distribution method.

8.3 Return of Voted Ballots

The return of voted ballots poses threats that are more serious and challenging than the threats to delivery of blank ballots and registration and ballot request. In particular, election officials must be able to ascertain that an electronically-returned voted ballot has come from a registered voter and that it has not been changed in transit. Because of this and other security-related issues, the threats to the return of voted ballots by e-mail and web are difficult to overcome.

Use of Telephone for Return of Voted Ballots:
Voting over the telephone presents a number of security challenges. Election officials would have to use methods other than voter signatures to authenticate voters; these methods, such as use of a PIN, which could be stolen, may present greater risks. Furthermore, a great deal of trust must be placed in the receiving site's equipment to accurately record votes, as there would be no opportunity for voters to directly verify that their ballots have been recorded correctly. The security challenges associated with telephone voting systems are difficult to mitigate using technology that is widely studied and deployed today.

Use of Fax for Return of Voted Ballots:
Faxing voted ballots to election officials presents some challenges for maintaining voter privacy and preventing the modification or destruction of voted ballots. Proper procedures may effectively mitigate these threats and reduce the overall risk to a manageable level.

Use of E-mail for Return of Voted Ballots:
The use of e-mail to return ballots presents several significant security challenges. Several different computer systems are involved in sending an e-mail from a voter to an election official. Many of these systems, such as the voters' computers and e-mail servers, are outside the control of election officials. Attacks on these systems could violate the privacy of voters, modify ballots, or disrupt communication with election officials. Because other individuals or organizations operate these systems, there is little election officials can do to prevent attacks on these systems. The security challenges associated with e-mail return of voted ballots are difficult to overcome using technology widely deployed today.

Use of Web for Return of Voted Ballots:
Casting ballots via the web poses a large number of security challenges that are difficult to overcome. Using this transmission method, voters would log into a web site and submit their selections on a web page. A great deal of trust must be placed in the software on the election server to accurately record votes, as there would be no opportunity for voters to directly verify that their ballots have been recorded correctly.

Furthermore, like e-mail voting systems, a web-based system for casting ballots would rely on computer systems outside the control of election officials. Attacks on these systems, such as voters' computers, could significantly threaten the integrity of elections or the ability of voters to cast ballots. Less sophisticated attacks, such as phishing and spoofing, could trick voters into giving up their voting credentials to an attacker. Such attacks are common in the banking industry, and difficult to defend against. There have been and continue to be significant problems in this industry. Technology that is widely deployed today is not able to mitigate many of the threats to casting ballots via the web.

8.4 Suggested Next Steps
The threat analysis documented in this paper identifies blank ballot distribution methods as a potential area to immediately improve UOCAVA voting without threatening the security of elections. Fax, e-mail and web-based systems could distribute blank ballots quickly and reliably to voters, significantly reducing the ballot delivery times faced by mailing ballots to voters and improving the UOCAVA voting experience for citizens overseas. In addition, registration and ballot requests can also take advantage of these distribution methods, but there are more threats when handling personal information from voters. Voted ballot return remains a more difficult issue to address, however emerging trends and developments in this area should continue to be studied and monitored.

A number of states already distribute blank ballots via fax or e-mail. However, at this time there are no guidelines that document best practices for fax, e-mail or web-based distribution of ballots. Developing such guidelines could help additional states develop methods for distributing ballots using these transmission methods, and potentially improve the procedures and technical controls already in place in the states currently using these systems.

References

[1] FIPS 199, Standards for Security Categorization of Federal Information and Information Systems, February 2004.

[2] NIST SP 800-30, Risk Management Guide for Information Systems, July 2002.

[3] NIST SP 800-53 Rev. 2, Recommended Security Controls for Federal Information Systems, December 2007.

[4] NIST SP 800-52, Guidelines for the Selection of Use of Transport Layer Security (TLS) Implementations, June 2005.

[5] NIST SP 800-63-1, Electronic Authentication Guidelines, February 2008 (Draft).

[6] Voting Assistance Guide, Federal Voting Assistance Program, 2008.

[7] Dierks, T. and Rescorla, E., *The TLS Protocol Version 1.2*, Internet Engineering Task Force, Request for Comment 5246, August 2008, http://tools.ietf.org/html/rfc5246

[8] ISO 32000-1:2008, Portable Document Format—Part 1: PDF 1.7.

[9] Klensin, J.. Simple Mail Transfer Protocol, Internet Engineering Task Force, Request for Comment 5321, October 2008, http://tools.ietf.org/html/rfc5321

[10] FIPS 140-3, Security Requirements for Cryptographic Modules, July 2007 (Draft).

[11] Rockapetris, P., Domain Names- Concepts and Facilities, Internet Engineering Task Force, Request for Comment 1034, October 1987, http://tools.ietf.org/html/rfc1034

[12] NIST SP 800-81, Secure Domain Name System (DNS) Deployment Guide, May 2006.

[13] Vulnerability Note VU#800113, US-CERT, July 2008, http://www.kb.cert.org/vuls/id/800113

[14] Jefferson, D., Rubin, A., Simons, B., Wagner, D., A security analysis of the Secure Electronic Registration and Voting Experiment (SERVE), January 2004, http://www.servesecurityreport.org

[15] Bonsor, K. and Strickland, J. How E-voting Works. http://people.howstuffworks.com/e-voting4.htm

[16] Federal Voting Assistance Program; Voting Over the Internet Pilot Project Assessment Report, Department of Defense, Washington Headquarters Services, June 2001.

[17] Independent Review Final Report for the Interim Voting Assistance System (IVAS), August 2006.

[18] Resources for Overseas Citizens and Military Voters. Election Assistance Commission. http://www.eac.gov/voter/overseas-citizens-and-military-voters

[19] Federal Voting Assistance Program, Department of Defense, http://www.fvap.gov

[20] The Uniformed and Overseas Citizens Absentee Voting Act, United States Department of Justice, Civil Rights Division, http://www.usdoj.gov/crt/voting/misc/activ_uoc.php

[21] Uniformed and Overseas Citizens Absentee Voting Act (UOCAVA), (as modified by the National Defense Authorization Act for FY 2005). http://www.fvap.gov/resources/media/uocavalaw.pdf

[22] S/MIME Working Group, Internet Engineering Task Force, http://www.imc.org/ietf-smime/

[23] FIPS 186-3, Digital Signature Standard (DSS), November 2008 (Draft).

[24] Military Postal Service Agency, http://hqdainet.army.mil/mpsa/

[25] USPS- Send International Mail, United States Postal Service, http://www.usps.com//international/sendmail.htm

[26] National Vulnerability Database, National Institute of Standards and Technology, http://nvd.nist.gov/

[27] Overseas Vote Foundation, https://www.overseasvotefoundation.org/

[28] Help America Vote Act of 2002, http://www.fec.gov/hava/law_ext.txt

Appendix: Acronyms

DNS	Domain Name System
DoD	Department of Defense
EAC	Election Assistance Commission
ETS	Electronic Transmission Service
FIPS	Federal Information Processing Standard
FPCA	Federal Post Card Application
FVAP	Federal Voting Assistance Program
FWAB	Federal Write-In Absentee Ballot
HAVA	Help America Vote Act of 2002
IVAS 2004	Interim Voting Assistance System
IVAS 2006	Integrated Voting Alternative Site
NIST	National Institute of Standards and Technology
PDF	Portable Document Format
PIN	Personal Identification Number
PKI	Public Key Infrastructure
PSTN	Public Switched Telephone Network
SERVE	Secure Electronic Registration and Voting Experiment
S/MIME	Secure/Multipurpose Internet Mail Extensions
SSL	Secure Socket Layer
TLS	Transport Layer Security
UOCAVA	Uniformed and Overseas Citizens Absentee Voting Act
VOI	Voting Over the Internet
VoIP	Voice Over Internet Protocol

www.ingramcontent.com/pod-product-compliance
Lightning Source LLC
Chambersburg PA
CBHW081734170526
45167CB00009B/3818